Lecture Notes in Earth Sciences 67

Editors:
S. Bhattacharji, Brooklyn
G. M. Friedman, Brooklyn and Troy
H. J. Neugebauer, Bonn
A. Seilacher, Tuebingen and Yale

W0043779

Springer-Verlag Berlin Heidelberg GmbH

Sharon L. Webb

Silicate Melts

With 33 Figures

 Springer

Author

Dr. Sharon L. Webb
Research School of Earth Sciences
Petrophysics Group
The Australian National University
Canberra ACT 0200, Australia

"For all Lecture Notes in Earth Sciences published till now please see final pages of
the book"

Cataloging-in-Publication data applied for

Die Deutsche Bibliothek - CIP-Einheitsaufnahme

Webb, Sharon L.:
Silicate melts / Sharon L. Webb.

(Lecture notes in earth sciences ; Vol. 67)
ISBN 978-3-540-63129-3   ISBN 978-3-540-69152-5 (eBook)
DOI 10.1007/978-3-540-69152-5

ISSN 0930-0317
ISBN 978-3-540-63129-3

Typesetting: Camera ready by author
SPIN: 10556605   32/3142-543210 - Printed on acid-free paper

Table of Contents

Symbols

α_v	—	coefficient of volume thermal expansion	K^{-1}
c	—	velocity	$m\,s^{-1}$
c_p	—	heat capacity	$J\,g^{-1}\,K^{-1}$
c_{pe}	—	relaxed heat capacity	$J\,g^{-1}\,K^{-1}$
c_{pg}	—	unrelaxed heat capacity	$J\,g^{-1}\,K^{-1}$
H	—	enthalpy	$J\,g^{-1}$
f	—	frequency	Hz
G^*	—	complex shear modulus	GPa
G'	—	real component of the shear modulus	GPa
G''	—	imaginary component of the shear modulus	GPa
G_∞	—	unrelaxed shear modulus	GPa
K^*	—	complex volume (bulk) modulus	GPa
K'	—	real component of the volume modulus	GPa
K''	—	imaginary component of the volume modulus	GPa
K_0	—	relaxed volume modulus	GPa
K_1	—	relaxation component of the volume modulus	GPa
K_∞	—	unrelaxed volume modulus	GPa
K_s	—	adiabatic volume modulus	GPa

$$K_s = \rho c^2 = \beta^{-1} = -V \cdot \left. \frac{dP}{dV} \right|_s$$

M	—	modulus	GPa
M	—	longitudinal modulus	GPa
P	—	physical property	
q	—	quench-rate	$K\,s^{-1}$
V	—	volume	m^3
V_e	—	relaxed volume	m^3
V_g	—	unrelaxed volume	m^3
T	—	temperature	K
T_f	—	fictive temperature	K
T_g	—	glass transition temperature	K
η	—	viscosity	Pa s
η^*	—	complex viscosity	Pa s
η'	—	real component of viscosity	Pa s
η''	—	imaginary component of viscosity	Pa s

η_l	—	longitudinal viscosity	Pa s
η_s	—	shear viscosity	Pa s
η_v	—	volume viscosity	Pa s
ρ	—	density	g cm^{-3}
ε	—	strain	
$\dot{\varepsilon}$	—	strain-rate	s^{-1}
σ	—	stress	kg m s^{-2}
β	—	compressibility	GPa^{-1}
τ	—	relaxation time	s
τ_l	—	longitudinal relaxation time	s
τ_s	—	shear relaxation time	s
τ_v	—	volume relaxation time	s
ω	—	angular frequency	rad s^{-1}

1 Introduction

A general equation of state for silicate melts is the ultimate goal of the study of their physical and thermal properties. Data on the physical properties of silicate melts at high-temperature/low-viscosity conditions and at temperatures just above the glass transition (at high viscosities) gives access to a wide range of information about the density, compressibility and viscosity of melts at different conditions; from lava flows, to magma bodies, to sinking or floating of crystals in magmas, to the possibility of movement of melt through rocks (see Fig. 1.1). Extensive investigation of the composition dependence of silicate melt densities, compressibilities and viscosities at superliquidus temperatures have yielded models for the equation of state of melts at high temperatures and one atmosphere pressure (Lange and Carmichael, 1990; Richet and Neuville, 1992). Despite the abundant high temperature data, extrapolation of these data to magmatic temperatures ~600 K lower than the measurement temperature is hampered by lack of knowledge of the temperature dependence of thermal expansion, viscosity and compressibility. The vicinity of the glass transition temperature to the magmatic temperatures makes it very difficult to observe the physical properties of silicate melts at these conditions. Extrapolation of the data in composition space is hampered by the limited composition range over which data exists (see Lange and Carmichael, 1990) and the knowledge that linear extrapolations are not applicable when the structure of the melt varies as a function of composition or temperature. There is presently only a limited amount of data on the physical properties of silicate melts at high temperatures and pressures (e.g. shock wave equation of state: Rigden et al., 1988; Miller et al., 1991; density: Agee and Walker, 1988; viscosity: Kushiro, 1976; Scarfe et al., 1987).

There is, however, an increasing amount of ultrasonic and shock wave data on the velocity of compressional waves propagating through silicate melts. The data in the literature have been plagued by a relaxation (equilibration) of the structure occurring at a seemingly unpredictable temperature (see Bockris and Kojonen, 1960; Laberge et al., 1973; Sato and Manghnani, 1985; Rigden et al., 1988). This structural relaxation has also been observed in high-viscosity volume measurements (Taniguchi, 1989). Rivers and Carmichael (1987) measured the sound-wave velocity for a large number of silicate melts and pointed out that the occurrence of the volume structural relaxation was predictable, as it could be related to the shear viscosity successfully by the Maxwell relationship. Using the

Maxwell relationship it is thus possible to design experiments to investigate the structural relaxation occurring in silicate melts using a variety of physical property measurement techniques; and further, to reinterpret data in the literature.

The question is: what part of the structure of the silicate melt is relaxing on the timescale of the ultrasonic, shock-wave and volume measurements; and how can this information be used to tell us more about the structure of melts and the temperature and composition dependence of their physical properties and structure. Given an understanding of the nature of the physical effects of the structural relaxation it will be possible to determine the density, compressibility and viscosity of melts at high viscosity conditions as a function of timescale of measurement — obtaining information both on the timescale of relaxation of various physical properties (viscosity, enthalpy, density, compressibility) and the melt structure and the relaxed and unrelaxed properties. The structural relaxation discussed here is the slowest relaxation process occurring in a silicate melt at a specific temperature. This relaxation therefore represents the glass transition. As discussed in Sect. 2.7 this structural relaxation is the lifetime of the Si-O bonds in the melt. No extended species (e.g. "polymers") exist on longer timescales than the lifetime of the fundamental Si-O bond at a specific temperature.

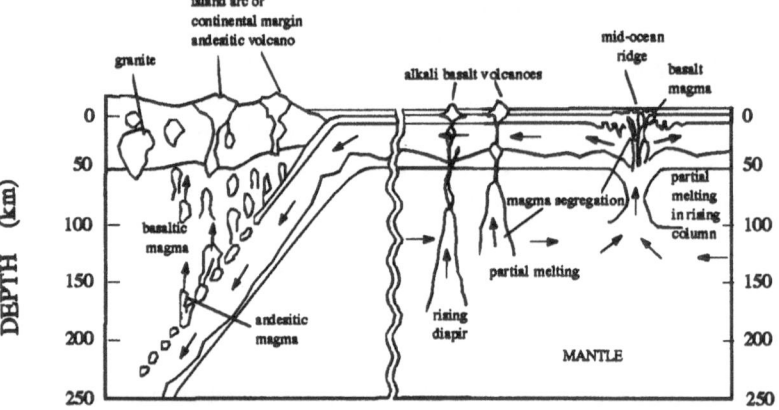

Fig. 1.1. A petrological model of melt beneath ocean ridges, ocean islands and continental/island arcs. (Redrawn after Ringwood, 1969.)

The following is a discussion of a range of different studies to determine the physical properties of silicate melts, based on a central theme of the structure of melts and the use of the structural relaxation phenomenon (especially the proposal that the same form of equation can be used to describe the relaxation of volume, shear and enthalpy if these three properties are a function of the same structural relaxation in the melt) to calculate the physical properties of a melt. For example the use of calorimetry data to calculate viscosity and volume and thermal expansion in silicate melts at temperatures just above the glass transition.

2 Relaxation

2.1 The Glass Transition

The response of a melt to some physical perturbation consists of an instantaneous response (e.g. the elastic strain due to an applied step in stress (Rosen, 1982); or the instantaneous change in enthalpy due to the application of a step in temperature (DeBolt et al., 1976)) and a delayed response (viscous flow, or the structural relaxation contribution to the change in enthalpy). If the melt behaviour is measured instantaneously upon the application of a perturbation, only the instantaneous component of the response is observed, and the unrelaxed, glass behaviour of a melt is determined. When the response of a melt is observed a long time — micro-seconds, minutes or years — after the application of the perturbation, only the delayed response is operative and the relaxed, liquid behaviour is observed. It is the timescale — or frequency — of observation that determines whether the liquid or glass behaviour of a silicate melt is observed. The intermediate region in temperature- and observation time-space between the glass and liquid behaviour is the glass transition region. The relaxation time is related to temperature via the melt viscosity (see Sect. 2.3 and Eqn. 2.1).

2.2 Glass Transition Temperature T_g

When a silicate glass is heated across the glass transition, a time-dependent response of its physical properties occurs. The unrelaxed, glassy response of the melt approaches equilibrium, liquid behaviour over a finite period of time. This gradual approach of the physical properties to their equilibrium values is due to structural relaxation. At low temperatures and short timescales, below the glass transition, the response of the melt to a perturbation is that of a frozen structure. At high temperatures and long timescales, above the glass transition, the changing structure (configuration) of the melt contributes to the observed behaviour. As the change in melt behaviour at the glass transition is generally difficult to observe, the temperature derivative of a property [e.g. heat capacity = Δenthalpy/ΔT;

thermal expansion = Δvolume/ΔT] across the glass transition region is used to determine the temperature at which the structure of the melt relaxes – the glass transition temperature T_g. Quantitative structural relaxation models have been constructed (Narayanaswamy, 1971; Moynihan et al., 1976a,b; Scherer, 1984) to reproduce the details of the time-dependent response of melt properties (e.g. density, refractive index, volume, enthalpy) in the glass transition interval. Although the timescales over which various physical properties of a melt relax are not necessarily the same (i.e. it is not necessarily the same structural change which influences the relaxation of each physical property), the timescales of volume and viscosity relaxation (Rekhson et al., 1971) and refractive index and enthalpy relaxation (Sasabe et al., 1977) and longitudinal and shear relaxation (Siewert, 1993; Siewert and Rosenhauer, 1994) in silicate melts have been found to be indistinguishable.

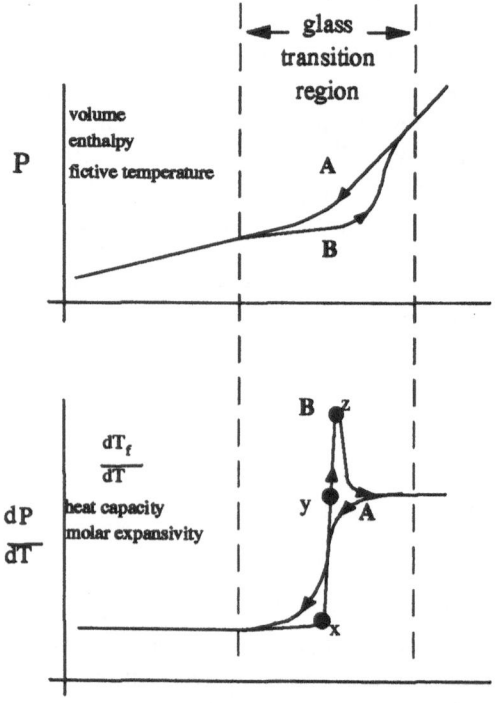

TEMPERATURE

Fig. 2.1. a Property P during cooling (A) and subsequent reheating (B) across the glass transition interval in temperature space. **b** The temperature derivative of the property P during cooling (A) and subsequent reheating (B) across the glass transition region.

Figure 2.1 shows a schematic plot of the behaviour of a physical property "P" of a glass forming melt and its temperature derivative for cooling and subsequent reheating at a constant rate. The temperature range during cooling in which the rate of the structural relaxation of the melt starts to become sufficiently slow that the property departs from its equilibrium value is the beginning of the glass transition region. The temperature range in which the structural relaxation becomes so slow that it effectively ceases on the experimental timescale marks the end of the glass transition region and the beginning of glass-like behaviour. Irrespective of the direction of temperature change, the direction of structural relaxation is always towards equilibrium. Consequently on reheating the glass, the property P and derivative curves follow paths different from those for the prior cooling. The glass transition temperature is defined as some characteristic temperature on the P or $\Delta P/\Delta T$ curve in the transition region during heating or cooling at a constant rate; for example the temperature of intersection of the extrapolated liquid and glass property curves during cooling, the extrapolated temperature of onset of the rapid increase in $\Delta P/\Delta T$ upon heating, or the temperature of inflection in the region of the rapid rise in the $\Delta P/\Delta T$ curve. T_g shows a dependence upon cooling- and heating-rate, with higher values of T_g being determined for faster cooling- and heating-rates. The dependence of T_g on cooling- and heating-rate therefore yields information on the kinetic parameter, the relaxation time (τ) of the melt structure.

2.3 The Maxwell Relationship

The relaxation timescale for shear deformation of a melt can be calculated by the Maxwell relationship (Maxwell, 1867)

$$\tau_s(T) = \frac{\eta_s(T)}{G_\infty(T)} \qquad (2.1)$$

where $\eta_s(T)$ is the relaxed Newtonian shear viscosity of the melt as a function of temperature and $G_\infty(T)$ is the unrelaxed elastic shear modulus of the melt. The unrelaxed shear modulus of silicate melts is essentially composition and temperature insensitive, and is taken to be 10 GPa ± 0.5 $\log_{10}$ units (Dingwell and Webb, 1990). The Maxwell relationship can be recast in terms of the volume modulus and volume viscosity in order to calculate the relaxation time of volumetric deformation of a melt (Gruber and Litovitz, 1964). The volume viscosity of silicate melts can be approximated by the shear viscosity (Dingwell and Webb, 1990) and the volume modulus of silicate melts is 30 GPa $\pm$ 0.5 $\log_{10}$ units (Bansal and Doremus, 1986; Rivers and Carmichael, 1987). Measured relaxation times for volume (τ_v) and shear (τ_s) deformation in melts have been found to be essentially the same as the calculated shear relaxation times, with $0.8 \ \tau_s \leq \tau_v \leq \tau_s$ (Tauke et al., 1968; B_2O_3 melt) and $\tau_s \approx \tau_v$ (Webb, 1991; $Na_2Si_2O_5$ melt). This suggests that the relaxation timescales for both shear and volume deformation in silicate melts are equivalent, and that physical properties affected by either of these structural relaxation processes will also have equivalent relaxation times. Given the large amount of viscosity data for silicate melts (e.g. Ryan and Blevins, 1987) the Maxwell relationship can be used to calculate the structural relaxation time curve for any melt composition in temperature-time space.

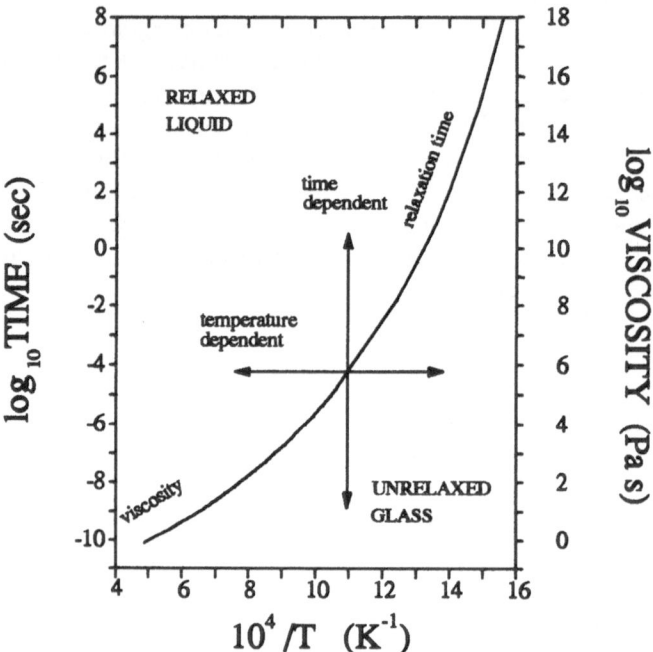

Fig. 2.2. The relaxation time curve for $Na_2Si_2O_5$ calculated from the Maxwell relationship. The relaxation time curve can be crossed at a constant temperature by varying the timescale of observation of the melt property. If the timescale of observation is held constant, the relaxation time curve can be crossed by varying the temperature of the melt.

The relaxation time curve for $Na_2Si_2O_5$ is illustrated in Fig. 2.2 using the shear viscosity data of Fontana and Plummer (1979) and an elastic shear modulus of 10 GPa. If the response of the melt to some perturbation is observed on a long timescale (i.e. a long time after the perturbation), at high temperatures, the relaxed liquid properties of the melt will be determined. If the behaviour is observed at short timescales (i.e. a short time after the perturbation) and low temperatures, the unrelaxed glassy properties of the melt will be measured. The relaxation time curve calculated from the Maxwell relationship then describes the glass transition temperature as a function of observation time. The silicate melt used in windows displays glassy properties at 300 K as its relaxation time is ~ 300 years (see Eqn. 2.1 and Table 2.1).

Table 2.1. Shear viscosity of some common liquids.

liquid	Temperature °C	η_s $\log_{10}$ Pa s	remarks
acetone	20	-3.5	extremely low viscosity
water	20	-3.0	-
glycol	20	-1.7	-
castor oil	20	0.0	low viscosity
glucose	40	8.5	high viscosity
glucose	20	13.0	-
window glass	20	20	-
Earth's mantle	~ 2000	20-21	extremely high viscosity

There are two ways in which to observe the changes in physical properties of a silicate melt across the glass transition (see Fig. 2.2). The first method is to hold the timescale of observation constant and to increase or decrease the temperature of the melt (e.g. scanning calorimetry or dilatometry at fixed heating-rates: observation time ~10-100 s; Archimedean density measurements: observation time ~1000 s; fixed frequency ultrasonic measurements: observation time ~5 ns). The timescale of observation fixes the glass transition in time-space, and also in viscosity-space through Eqn. 2.1. Increasing or decreasing the timescale of observation will decrease or increase, respectively, the observed glass transition temperature. For observation timescales ranging from 5 ns to 1000 s, the viscosity at the glass transition temperature ranges from ~1.7 to ~13 $\log_{10}$ Pa s for silicate melts in general, and the glass transition temperature for $Na_2Si_2O_5$ ranges from 1353 to 698 K.

The second method is to hold the temperature constant and to vary the timescale of observation of the melt properties (e.g. variable frequency ultrasonic measurements: observation time ~1-50 ns; "a.c.". calorimetry: observation time 8 μs-0.4 s; viscosity measurements: observation time 10^{-3}-10^5 s). As the temperature remains constant using this second technique, the observed variation in melt properties in the vicinity of the glass transition is not influenced by small temperature dependent changes occurring in the structure of the melt, but is a true measurement of the time dependent response of the melt to some perturbation. Increasing or decreasing the temperature at which the melt behaviour is observed will decrease or increase respectively, the observation timescale necessary to observe the glass transition. For a chosen temperature, a timescale of observation can be found such that $T_{chosen} = T_g$. Fixing the glass transition temperature also fixes the viscosity at which the glass transition occurs. For temperatures in the range 1538 to 733 K ($Na_2Si_2O_5$ viscosities from 1 to 11 $\log_{10}$ Pa s respectively), observation timescales of 1 ns (ultrasonic timescales) to 10 s (calorimetric timescales) are required in order to match T_g with the chosen temperature. The industrial definition of T_g is the temperature at which the shear viscosity of the melt is 10^{12} Pa s (Illig, 1991). This empirical definition has been successful as most industrial measurements are performed on a measurement timescale of 10-100 s. Using this timescale as the relaxation time, the Maxwell relationship calculates a viscosity of 10^{11}-10^{12} Pa s, as the viscosity at the glass transition imposed by the observation time.

2.4 Thermorheological Simplicity

Thermorheological simplicity (TRS) is derived from the principle of temperature (actual and fictive) – time equivalence. Thermorheological simplicity is a useful first approximation to the behaviour of a melt in the glass transition region. The application of thermorheological simplicity to a set of data is strictly valid only if the shape of the relaxation time function remains constant as a function of temperature (i.e. changes in temperature merely shift the relaxation function along the $\log_{10}$ time axis without altering its shape (Narayanaswamy, 1988)). There exist very few measurements which can test the applicability of thermorheological simplicity for silicate melts across more than 100 K in temperature. The data of Macedo et al. (1968), Simmons and Macedo (1970) and Webb (1992a,b), however, would appear to indicate that thermorheological simplicity is not strictly true across a temperature range greater than 300 K, which is equivalent to a viscosity range of ~5 $\log_{10}$ Pa s for silicate melts.

2.5 Fictive Temperature T_f

The physical properties of a silicate melt determined at a fixed rate of observation as a function of temperature depend upon both the ambient temperature (T) and the structure of the melt (and also on the fixed observation timescale). Glasses quenched from silicate liquids preserve a structure that can be approximated to the equilibrium structure of the liquid at some fictive temperature T_f. To describe the unrelaxed (and relaxed) properties of a silicate melt it is necessary to specify the temperature and the fictive temperature of the melt (i.e. structure). For a liquid, the structure is in equilibrium and therefore $T_f = T$ (see Fig. 2.3). Upon cooling of the liquid into the glass transition region, the structure of the melt begins to deviate from equilibrium and $T_f > T$. With decreasing temperature T_f becomes temperature independent and describes the frozen structure of the glass. Upon reheating through the glass transition interval, the value of T_f gradually assumes that of the ambient temperature, and liquid values of melt properties are observed. Due to the finite rate of equilibration available for structural relaxation at the onset of the glass transition region, T_f lags behind T across a temperature range of ~100 K — the glass transition region — after which $T_f = T$ (DeBolt et al., 1976). The temperature derivative of T_f is zero at low temperatures where T_f is constant; is equal to one at high temperatures where $T_f = T$; and exhibits a maximum in the glass transition region upon heating as T_f must change at a faster rate than the heating-rate in order for the structure to catch up to that of the ambient temperature. The position of the peak in the $\Delta T_f / \Delta T$ curve is a function of the cooling- and heating-rate of the glass (Wong and Angell, 1976; Brawer, 1985). If the peak temperature is taken to represent the glass transition temperature T_g, this transition temperature is also dependent on the cooling- and heating-rate of the measurement.

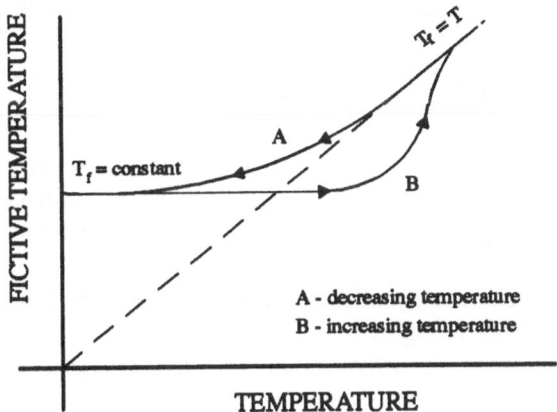

Fig. 2.3. The path of fictive temperature during cooling (A) and reheating (B) through the glass transition interval.

As the fictive temperature T_f describes the structure of the melt, and the relaxation of the melt structure effects the relaxation of melt properties, the curves in Fig. 2.1 represent the temperature dependence of any physical property of a silicate melt e.g. enthalpy (heat capacity = $\Delta H/\Delta T$), volume (thermal expansivity = $\Delta V/\Delta T$), strain due to a constant applied stress (modulus) or strain-rate due to a constant applied stress (viscosity).

2.6 Linear Viscoelastic Rheology

In the simplified case melts are considered to display Newtonian rheology and crystalline materials are considered to display elastic stress-strain behaviour. Investigation of the physical properties of silicate melts reveals that these materials can be described by viscoelastic rheology. In the case of ideal elasticity, Hooke's law defines the linear relation between stress (σ) and strain (ε) as

$$\sigma = M\varepsilon \tag{2.2}$$

for the modulus M. For Newtonian viscosity, the linear relationship between stress and strain-rate is

$$\sigma = \eta\dot{\varepsilon} \tag{2.3}$$

for viscosity η. Ideal linear elasticity is defined by three conditions (Nowick and Berry, 1972):
(i)   for every stress there is a unique equilibrium value of strain and vice versa (an important corollary is complete recovery)
(ii) the equilibrium response is achieved instantaneously
(iii) the stress-strain relationship is linear (Twice the stress results in twice the strain. This is the Boltzmann superposition principle; which requires that stresses resulting from individual strain increments add linearly (Rosen, 1982)).
Lifting the condition that the equilibrium response is achieved instantaneously results in condition (ii) becoming;
(ii) the equilibrium response is achieved only after the passage of some time and describes linear anelastic behaviour.
Lifting condition (i) requiring complete recoverability of the deformation describes linear viscoelastic behaviour (see Webb and Dingwell, 1995).

If a sinusoidal pressure oscillation is applied to a viscoelastic material, the bulk (volume) modulus is described by a complex modulus (Herzfeld and Litovitz, 1959):

$$K^*(\omega) = K_0 + \frac{K_1\omega^2\tau_v^2}{1+\omega^2\tau_v^2} + i\,\frac{K_1\omega\tau_v}{1+\omega^2\tau_v^2} \qquad (2.4)$$

where $\omega = 2\pi f$ is the angular frequency of the stress wave. K_0 is the relaxed modulus of the material and $K_\infty = K_0 + K_1$ is the unrelaxed modulus and τ_v is the volume relaxation time of the material. The volume relaxation time is related to the volume viscosity (Herzfeld and Litovitz, 1959; Gruber and Litovitz, 1964);

$$\tau_v = \frac{\eta_v}{K_1} \qquad (2.5)$$

where K_1 is the relaxational part of the volume modulus. The complex shear modulus is

$$G^*(\omega) = \frac{G_\infty\omega^2\tau_s^2}{1+\omega^2\tau_s^2} + i\,\frac{G_\infty\omega\tau_s}{1+\omega^2\tau_s^2}. \qquad (2.6)$$

The ratio of shear viscosity to shear modulus gives the timescale of shear relaxation of a melt:

$$\tau_s = \frac{\eta_s}{G_\infty}. \qquad (2.7)$$

The complex longitudinal modulus is

$$M^*(\omega) = M' + iM'' = K^*(\omega) + \tfrac{4}{3}G^*(\omega) \qquad (2.8)$$

and the complex longitudinal viscosity is

$$\eta_l^*(\omega) = \eta_l' + i\eta_l'' = \eta_v^*(\omega) + \tfrac{4}{3}\eta_s^*(\omega) \qquad (2.9)$$

as a function of frequency. Figure 2.4 illustrates the expected behaviour of viscosity and modulus of a viscoelastic material as a function of frequency in the vicinity of the relaxation time τ (with frequency ω increasing to the right). Similar behaviour is expected in the time domain, with observation time increasing to the left in Fig. 2.4.

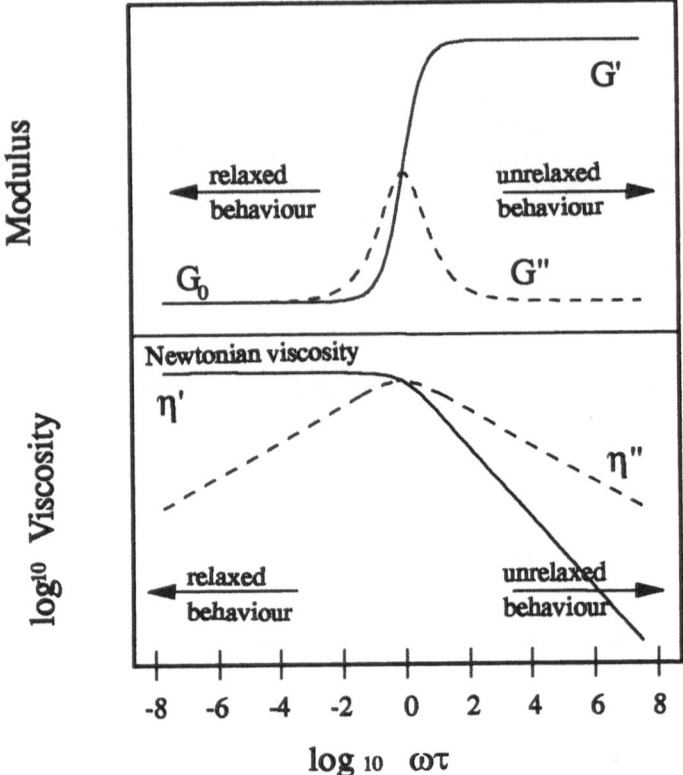

Fig. 2.4. Calculated frequency-dependent behaviour of longitudinal, volume and shear viscosities (η_l, η_v and η_s respectively) and moduli (M, K and G respectively) of a viscoelastic material with a relaxation time τ, plotted as a function of $\log_{10} \omega\tau$. The subscripts "0" and "∞" indicate zero frequency and infinite frequency values.

2.7 Structural Relaxation

The changes in physical properties of $Na_2Si_2O_5$ melt have been determined across the glass transition as functions of both temperature and timescale of measurement. The relaxation times and glass transition temperatures have been determined as functions of temperature and observation timescale, respectively. The temperature-dependent measurements were differential scanning calorimetry determination of the heat capacity and scanning dilatometric determination of the volume thermal expansion of the melt (Webb, 1992a). The time-dependent measurements were variable frequency ultrasonic determination of shear and volume moduli (Webb, 1991), variable frequency forced oscillation determination of the shear modulus (Mills, 1974) and fibre elongation determinations of shear viscosity (Webb and Dingwell, 1990a,b) done at fixed temperatures. The scanning determinations of relaxation processes differ from the observations performed at constant temperature and variable observation-rate in the latter are performed on materials in structural equilibrium, while the former are not in structural equilibrium. The relaxation times and glass transition temperatures are plotted in Fig. 2.5 together with the calculated Maxwell shear relaxation time.

The volume relaxation of $Na_2Si_2O_5$ was determined using dilatometric techniques (Webb et al., 1992) at a heating-rate of 5 K min^{-1} on a glass that had been cooled from its relaxed state at a rate of 5 K min^{-1}. The temperature at which the peak occurs is 752±2 K, with the onset of volume relaxation occurring at 720 K, for an observation timescale of 12 s. The timescale of volume relaxation for $Na_2Si_2O_5$ has also been determined using variable frequency ultrasonic techniques (Webb, 1991). Over an observation timescale range of 1 to 30 ns and a temperature range 1233 to 1443 K, the relaxation timescale was found to be equivalent to that calculated from the Maxwell relationship using the Newtonian shear viscosity and a shear modulus of 10 GPa. The observation time/temperature regions in which the volume relaxation was observed in these measurements is indicated in Fig. 2.5.

The enthalpy relaxation of $Na_2Si_2O_5$ has been observed at a heating-rate of 5 K min^{-1} (equivalent to an observation time of 12 s) on a glass quenched from a liquid at 5 K min^{-1} using calorimetric techniques (Webb et al., 1992). The peak in the heat capacity curve occurs at a temperature of 748±2 K, with the deviation from glassy to liquid behaviour occurring over a temperature range from 720 to 767 K. This relaxation range is plotted on the relaxation time curve for $Na_2Si_2O_5$ in Fig. 2.5

The relaxation timescale of shear deformation of $Na_2Si_2O_5$ has been determined using fibre elongation techniques (Webb and Dingwell, 1990b) at a temperature of 720 K and an observation timescale of 2-2000 μs. At long timescales of observation of the melt, a linear relationship between strain-rate and stress was observed and a Newtonian (linear stress-strain rate behaviour) viscosity was determined. With increasing strain-rate (decreasing timescale of observation), as the timescale of observation approaches the relaxation time of the melt, the stress-strain rate response falls out of equilibrium and the observed viscosity decreases. The temperature-timescale range in which the onset of shear relaxation was observed is plotted in Fig. 2.5. Variable frequency ultrasonic techniques have also been used to determine the timescale of shear relaxation in $Na_2Si_2O_5$ for a temperature range 1233 to 1443 K and observation timescale range of 1 to 20 ns (Webb, 1991). The relaxation time for shear deformation at these temperatures is that calculated from the Maxwell relationship. The shear relaxation timescale for $Na_2Si_2O_5$ has been determined by Mills (1974) in a temperature range 687-785 K using forced torsion oscillations in the frequency range 1 mrad s^{-1} < ω < 1 rad s^{-1}.

The timescale- temperature range in which frequency dependent shear modulus was observed is illustrated in Fig. 2.5.

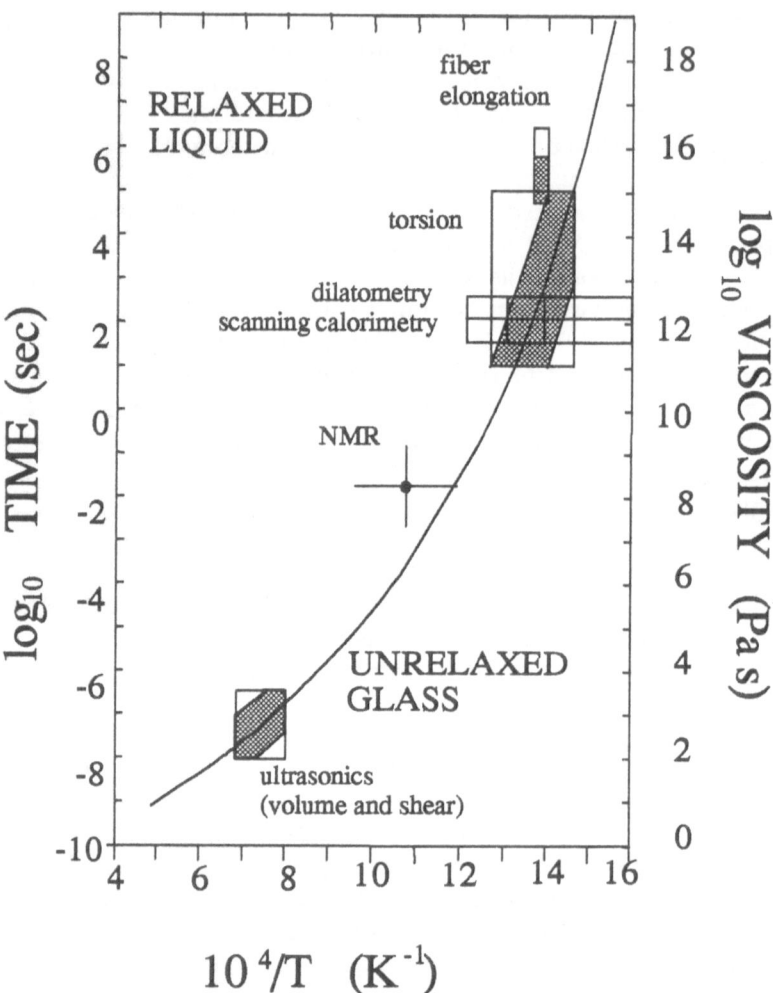

Fig. 2.5. The relaxation time curve for $Na_2Si_2O_5$ divides timescale-temperature space into regions of relaxed and unrelaxed behaviour. The rectangles illustrate the timescale-temperature regimes in which a variety of techniques were used to determine melt rheology. The hatched areas indicate the timescale-temperature regimes in which shear (fibre elongation, torsion, ultrasonics), volume (dilatometry, ultrasonics) and enthalpy (scanning calorimetry) relaxation have been observed. (redrawn from Webb, 1992a).

The timescale on which oxygens in $Na_2Si_2O_5$ exchange between bridging and non-bridging sites has been determined by Liu et al. (1988) using NMR techniques. This timescale has also been plotted on the relaxation timescale versus temperature plot for $Na_2Si_2O_5$ (Fig. 2.5).

The response of a melt to each physical perturbation will have its own relaxation (equilibration) timescale. Different structural components may also respond to the same perturbation on differing timescales. Therefore a glass transition may be described for every perturbation and its associated physical property (e.g. perturbation in stress and temperature, result in changes in modulus and enthalpy, respectively). The Maxwell relationship is seen in Fig. 2.5 to calculate the timescale of shear, volume and enthalpy relaxation in $Na_2Si_2O_5$ over a 750 K temperature range and an observation time range of 1 ns to 2000 s. At a viscosity of $\sim 10^{11}$ Pa s, the enthalpy relaxation time is the same as the volume relaxation time determined from dilatometry measurements and the shear relaxation time determined from shear deformation measurements. At a viscosity of 100 Pa s, the relaxation timescales for shear and volume deformation are equivalent.

The relaxation timescale calculated for a viscosity of 10^5 Pa s (923 K) is within error of the exchange timescale of Si-O bonds determined in $Na_2Si_2O_5$ by Liu et al. (1988). This suggests that the structural relaxation responsible for the relaxation of the properties volume, shear and enthalpy observed here is the making and breaking of Si-O bonds in the melt. Further work by Farnan and Stebbins (1990, 1994) on K_2O-SiO_2 melts found the activation energy for Si-O exchange to be identical to that for viscous flow, with the calculated Maxwell shear relaxation time and the exchange timescale of Si-O bonds being within 0.5 log_{10} units of each other.). Similar results have been obtained on Na_2O-SiO_2 melts by Stebbins et al. (1995) (see also Stebbins, 1995).

The relaxation time curve illustrates the observation timescale dependence of the glass transition. In the case of the long timescale of observation of the relaxation of shear and volume and enthalpy, the glass transition temperature is ~ 723 K. In the short observation time used in the ultrasonic study, the glass transition temperature increased to 1423 K. If the structural relaxation which results in the relaxation of viscosity, enthalpy, and modulus is the relaxation of Si-O bonds, the relaxation time curve describes the glass transition temperature for Si-O bond exchange. There also exists sodium ion motion occurring in $Na_2Si_2O_5$. The physical properties of the melt which are affected by this structural change will exhibit relaxation times equivalent to the timescale on which these ions move. This is illustrated for the case of Na_2O-$3SiO_2$ melt in Fig. 2.6. Dingwell and Webb (1990) have drawn the relaxation map for this melt as a function of temperature, with both the Si-O relaxation and the Na relaxation being indicated.

As the relaxation of all of the physical processes discussed above are the first relaxation observed as a function of decreasing observation timescale, or decreasing temperature, we are discussing the slowest relaxation of structure in these silicate melts. The slowest structural relaxation is called the α relaxation

(Wong and Angell, 1976). Faster relaxation processes occurring in the melt at the same temperature as the slowest, are labelled the β, γ etc. relaxation. The glass transition represents the slowest relaxation process observed at a fixed temperature in a silicate melt. Above the glass transition (at longer times of observation and high temperatures) no extended species (e.g. polymers) exist on a longer timescale than the lifetime of the fundamental Si-O bond.

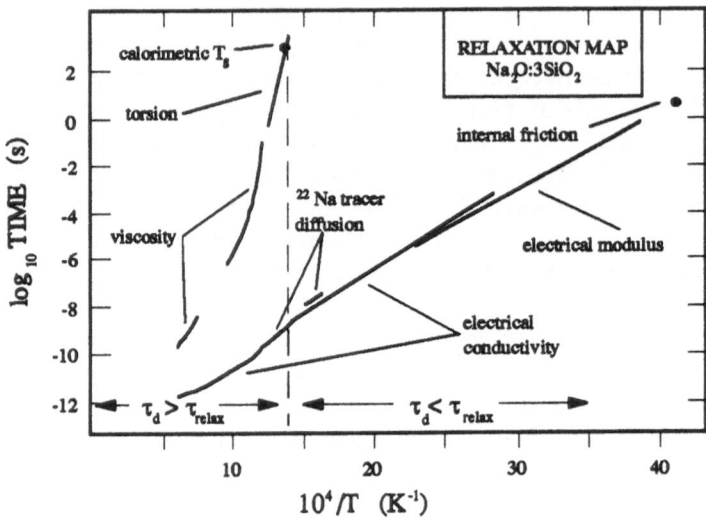

Fig. 2.6. Relaxation timescales in Na_2O-$3SiO_2$ melt. Structural relaxation data from the torsional study of Mills (1974) and the viscosity studies of Poole (1948) and Bockris et al. (1955). Alkali and electrical relaxation data from the tracer diffusion work of Johnson et a. (1951), the electrical modulus data of Provencano et al. (1972), the conductivity studies of Babcock (1934) and Seddon et al. (1932) and the internal friction data of Day and Steinkamp (1971). Calorimetric T_g data from Richet and Bottinga (1984). The inflection in electrical and diffusion data at T_g result from the experimental duration approaching the structural relaxation timescale. (redrawn after Dingwell, 1990)

The following discussion is confined to the slowest structural relaxation in silicate melts – the exchange of Si-O bonds. It is assumed, and data is provided to substantiate that the Maxwell relationship can be used to calculate the relaxation timescale for volume, shear and enthalpy relaxation and that the observed relaxation of the physical properties volume and shear moduli and viscosities, and enthalpy are due to the same structural relaxation process within the melt; and that this is the relaxation of the Si-O bonds in the melt structure.

3 Shear Relaxation

Given the general applicability of the Maxwell relationship to calculate the timescale of structural relaxation, the deformation conditions under which shear and volume relaxation are to be expected in silicate melts can be estimated. In the case of shear relaxation, especially, the question of non-Newtonian rheology arises. It has been generally assumed, for lack of evidence to the contrary, that all silicate melts are Newtonian, that is there is a linear relationship between stress and strain-rate. Inasmuch as the structure of a melt has a relaxation (equilibration) time that is a function of temperature, if the melt is sheared at a rate faster than the relaxation rate, the effects of structural relaxation should be observed and the stress necessary to achieve the deformation rate should decrease (see Sect. 2.6).

3.1 Time- and Frequency-Domain Measurements

A melt of shear viscosity η_s has a relaxation time of η_s/G_∞, or a relaxation strain-rate of $\dot\varepsilon_r = (\eta_s/G_\infty)^{-1}$. For strain-rates faster than the relaxation strain-rate non-linear stress-strain rate behaviour — that is non-Newtonian rheology — is expected to occur in silicate melts. The shear viscosity of a melt can be determined in time-space by the application of a constant stress and the measurement of the resulting strain-rate. In order to reach the relaxation strain-rate, a stress of $\eta_s\dot\varepsilon_r$ must be achieved (based on Newtonian viscosity calculations). For a shear viscosity of 1 Pa s, a strain-rate of $\sim 10^{10}$ s^{-1} and a stress of $\sim 10^{10}$ N m^{-2} is necessary. For a melt of Newtonian shear viscosity 10^{11} Pa s a strain-rate of $\sim 10^{-1} s^{-1}$ and a stress of $\sim 10^{10}$ N m^{-2} is required to match this condition (as $G_\infty \cong 10$ GPa, the stress required to reach the relaxation strain-rate is also 10 GPa [1 Pa = 1 N m^{-2}]). The calculated onset on non-Newtonian shear viscosity should occur at strain-rates ~ 2 orders of magnitude less than the relaxation strain-rate (Herzfeld and Litovitz, 1959; Simmons et al., 1982). Most techniques commonly used to determine shear viscosity as a function of observation time (see Zarzycki, 1991; Dingwell et al., 1993a) cannot reach the extreme conditions of strain-rate, which explains why very few observations of non-Newtonian viscosity have been documented.

Shear relaxation can also be measured in the frequency domain. The relaxation of strain can be observed directly by applying a sinusoidal shear stress and observing the phase shift between the applied sinusoidal shear stress and the resulting sinusoidal shear strain. This method involves the application of stresses of approximately 100 kPa. Shear relaxation can be observed in an indirect manner by determining the velocity of a shear wave traveling through the melt as a function of the frequency of the sinusoidal deformation. This involves stresses of ~kPa and strains of ~10^{-8}.

3.2 Fibre Elongation

In the case of fibre elongation measurements, in which the melt fibre being deformed has a diameter of 0.2 mm, it is possible to apply a stress of 100 MPa and to achieve the strain-rate necessary to observe non-Newtonian stress-strain rate behaviour. In the case of a melt of rhyolitic composition, the existence of Newtonian viscosity using this technique was established over 4 orders of magnitude of strain-rate (Webb and Dingwell, 1990a) from 10^{-7} to $10^{-3.5}$ s^{-1} and stresses of $10^{4.5}$ Pa to 10^8 Pa s for a viscosity of $10^{11.2}$ Pa s. Upon the strain-rate approaching within 2 orders of magnitude of the calculated relaxation strain-rate, there was an increase in strain-rate for a fixed stress and a non-linear stress-strain-rate behaviour occurred. Such non-Newtonian viscosity has been observed for 11 silicate melt compositions (see Fig. 3.1) in both simple binary systems and also more complex oxide melts ranging from 50-80 mol% SiO_2 (Webb and Dingwell, 1990a,b). Similar behaviour has been observed by Li and Uhlmann (1970) and Simmons et al. (1982) in Rb_2O-SiO_2 and Na_2O-CaO-SiO_2 melts. It is possible that this decrease in viscosity could be caused by viscous heating of the melt. An increase in temperature of 8 K is required to cause the observed decrease in viscosity, and the maximal temperature increase in the fibre is <0.1 K, based on the emissivity and surface area and volume of the fibre (Li and Uhlmann, 1970; Webb and Dingwell, 1990a,b).

As this onset of non-Newtonian viscosity results in a run-away deformation of the melt, and there is only a limited amount of vertical space in the furnace, viscosity decreases of 1.5-3.0 log$_{10}$ Pa s are observed before the end of the measurement. The accuracy of the observed non-Newtonian viscosity is not sufficient to discuss the composition dependence of the stress-strain rate behaviour of the melts. The small temperature range over which data is available from the study of Webb and Dingwell (1990b) for $Na_2Si_4O_9$ melt is not large enough to resolve any temperature dependence of the onset of the relaxation strain-rate.

Normalisation of the experimental strain-rates to the calculated relaxation strain-rate eliminates the composition and temperature dependence of the onset of non-Newtonian behaviour. Webb and Dingwell (1990b) have shown for a range of silicate melt compositions that the onset of non-Newtonian behaviour occurs for strain-rates 3 orders of magnitude less than the calculated Maxwell relaxation strain-rate. The onset of non-Newtonian behaviour occurs at strain-rates less than expected for a single relaxation time and a single relaxing structure. Most studies of the relaxation behaviour of silicate melts at high viscosities observe behaviour which is consistent with the relaxation of the melt structure having a distribution of relaxation times. Any distribution in relaxation times would lead to a broadening of the relaxation zone and the occurrence of non-relaxed behaviour (non-Newtonian rheology) at lower strain-rates than those predicted from the single relaxation time theory employed in the derivation of the Maxwell relationship.

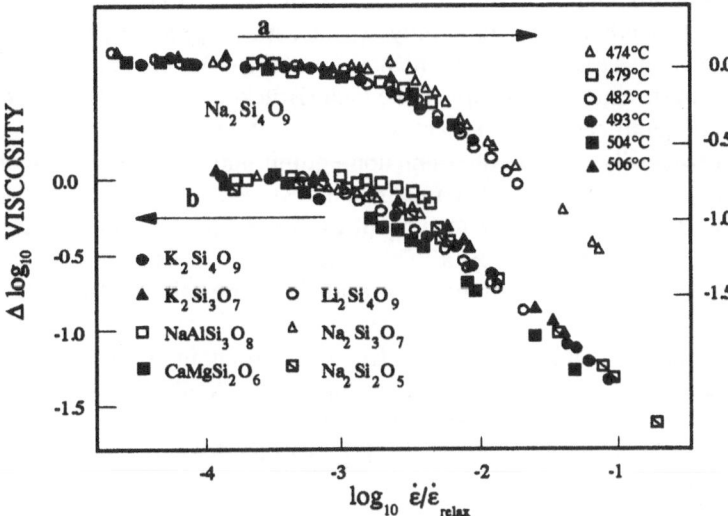

Fig. 3.1. a A reduced plot of viscosity (relative to Newtonian) versus strain-rate (relative to relaxation strain-rate) for $Na_2Si_4O_9$ at temperatures of 747, 752, 755, 777 and 779 K. **b** A reduced plot of viscosity versus strain-rate for the alkali-silicates, albite and diopside melts. These reduced plots, based on the Maxwell equation remove the temperature and composition dependence of the non-Newtonian viscosity of silicate melts. (redrawn from Webb and Dingwell, 1990b)

The normalised viscosity versus strain-rate curves for these melts do not show any difference in relaxation time distribution as a function of composition. The distribution of relaxation in the viscoelastic region therefore appears to be unrelated to the melt chemistry. This observation is a valuable simplification of the onset of non-Newtonian rheology and a useful approximation for modeling igneous processes. The accuracy of these stress-strain rate measurements in a dilatometer, especially in the run-away phase of deformation is low (± 0.3 log_{10} Pa s). Therefore it is possible that the subtleties of the composition dependence of the relaxation time distribution in relaxation are lost. This technique for measuring the onset of non-Newtonian behaviour involves the application of very large stresses and it is also possible that the deformation in the run-away phase is no longer described by linear stress-strain behaviour.

3.2.1 Brittle failure

This type of measurement underlines an important distinction to be made between low strain measurements of the glass transition (e.g. torsion, ultrasonic wave propagation) and high strain measurements such as fibre elongation. In fibre elongation type measurements, the stress required to approach the relaxation strain-rate approximates the tensile strength of the melt and the material fails before the structure of the melt is fully relaxed. It is expected that the large non-equilibrium strains associated with the non-Newtonian flow in the fibre will result in significant structural anisotropy in the melt (Brückner, 1987). If these structural changes are significant with respect to viscosity the two types of measurement may not necessarily result in the same non-equilibrium properties being observed in the non-Newtonian region. The final result observed in the studies of Webb and Dingwell (1990a,b) due to the loading of the fibre was the brittle failure of the fibre at viscosity of $\sim 10^{12}$ Pa s. This observation illustrates that in natural processes, stresses sufficiently high to result in brittle failure of a melt are also capable of producing non-Newtonian rheology. This observation would not be possible from measurements in the frequency domain. It is clear from the observation of volcanic ash that brittle failure of the rhyolitic melt occurs during eruption. This leads to the suggestion that viscous flow of melts during eruptions involving brittle failure of the magma will pass through a stage of unrelaxed, non-Newtonian deformation.

3.3 Forced Torsion

The composition dependence of non-Newtonian rheology can be observed using techniques which have higher accuracy in the measurements of stress and strain; and which allow measurement of the non-Newtonian behaviour across the entire viscoelastic region, resulting in the calculation of a relaxation time distribution function for each melt composition.

Forced torsion oscillation is such a method of observing strain-rate dependent non-Newtonian behaviour in melts as a function of strain-rate, which allows us to observe the composition effects on this behaviour. Webb (1992b), Bagdassarov et al. (1993, 1994) have applied a sinusoidally oscillating shear stress to cylinders of melts with a diameter of 8 mm and length 30 mm. The frequency of the shear strain-rate can be varied by 4 orders of magnitude from 0.005 to 10 Hz ($\omega = 2\pi f =$ 0.03-63 rad s^{-1}). The applied sinusoidal torque is 1.5-2.5×10^{-3} N m and the resulting angle of twist in the melt is ~10^{-5} rad. As this is a frequency measurement, there are two components of deformation in the melt. There is the deformation in-phase with the applied stress and the deformation 90° out-of - phase with the applied stress. These are the instantaneous recoverable elastic deformation, and the time dependent recoverable, and non-recoverable viscous deformation of the melt, respectively. The amplitude of the in- and out-of-phase components of deformation vary as a function of the frequency of the applied stress, with the out-of-phase (imaginary) component increasing in the vicinity of the relaxation frequency.

The relaxed and the unrelaxed shear viscosity and shear modulus of a range of silicate melts have been determined by forced oscillation (see Mills, 1974; Perez et al., 1981; Webb, 1992b; Bagdassarov et al., 1993, 1994). The shear viscosity is seen to become non-Newtonian with increasing strain-rate (increasing frequency of the applied deformation). This strain-rate dependence of the shear viscosity is similar to the behaviour observed in the fibre elongation measurements. The onset of non-Newtonian viscosity is however observed within 2 orders of magnitude of the relaxation strain-rate indicating that indeed the fibre elongation measurements are no longer within the range of linear stress-strain physics. In contrast to the fibre elongation measurements, this technique does indeed result in the observation of composition dependent distribution of the relaxation spectra (e.g. Bagdassarov et al., 1993).

As this is a frequency domain measurement, the presence of both elastic and viscous deformation in the melt results in deformation in phase with the applied stress — the real part of the shear modulus and deformation 90° out of phase with the applied stress — the imaginary part of the shear modulus. The frequency dependent shear modulus is then

$$G^*(\omega) = G^*(\omega)\cos[\psi(\omega)] + i\,G^*(\omega)\sin[\psi(\omega)] = G'(\omega) + i\,G''(\omega) \qquad (3.1)$$

where ω is the angular frequency of the applied stress and G' and G'' are the real and imaginary components of the shear modulus respectively. The complex frequency dependent shear viscosity can then be calculated from the shear modulus (Webb, 1991)

$$\eta_s^*(\omega) = \frac{G^*(\omega)}{i\omega} = \frac{G''(\omega)}{i\omega} - i\,\frac{G'(\omega)}{i\omega} = \eta_s'(\omega) + i\,\eta_s''(\omega) \qquad (3.2)$$

The real and imaginary components of the shear modulus of a bubble- and crystal-free rhyolite composition melt determined over a temperature range 973-1173 K are presented in Fig. 3.2 as a function of the normalised frequency $\omega\tau_s$ (Webb, 1992b). The relaxation time τ_s was calculated at each temperature from

Eqn. 2.1 using the measured shear viscosity of Webb and Dingwell (1990a) for this melt composition. The infinite frequency shear modulus was found to be 30.5 ±2.5 GPa. The dashed line in Fig. 3.2, is that calculated for a single relaxation time, where the relationship of Herzfeld and Litovitz (1959) is used to describe the expected relaxation behaviour for a single relaxation time τ_s:

$$G^*(\omega) = \frac{G_\infty\omega^2\tau_s^2}{1+\omega^2\tau_s^2} + i\frac{G_\infty\omega\tau_s}{1+\omega^2\tau_s^2} = G'(\omega) + i\,G''(\omega) . \qquad (3.3)$$

The solid curve is calculated by fitting a sum of the form;

$$G^*(\omega,\tau_s) = \sum_j a_j\, G^*\!\left(\omega,\, 10^{\left[\log_{10}(\tau_s)+j\right]}\right) \qquad (3.4)$$

to the real and imaginary parts of the data.

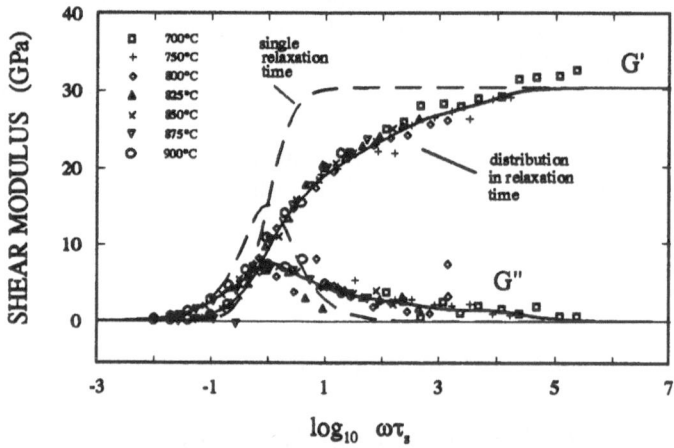

Fig. 3.2. The frequency-dependent real and imaginary components of the shear modulus of rhyolite composition melt from 973-1173 K. The solid line is fit to Eqn. 3.4. The dashed line is that calculated for a single relaxation time. The frequency is normalised to the Maxwell relaxation time. (redrawn after Webb, 1992b)

The calculated real and imaginary parts of the shear viscosity are presented in Fig. 3.3. The viscosity is normalised to the Newtonian shear viscosity for this melt calculated at each temperature. At low frequencies of deformation the real part of the viscosity is frequency independent and equal to the Newtonian shear viscosity determined by micropenetration viscometry on this sample. At high frequencies, the frequency dependent shear viscosity is 5 $\log_{10}$ Pa s less than the Newtonian shear viscosity.

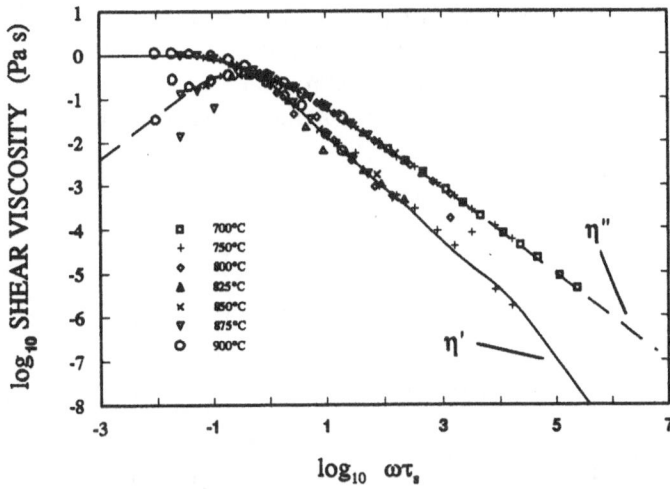

Fig. 3.3. The frequency-dependent real and imaginary components of the shear viscosity of rhyolite composition melt from 973-1173 K. The viscosity is normalised to the Newtonian (strain-rate independent) viscosity determined from micropenetration measurements. The frequency is normalised to the Maxwell relaxation time. The lines are fit to Eqn. 3.4 incorporating the distribution of relaxation times in Fig. 3.4. (redrawn after Webb, 1992b)

At these high viscosities, a distribution in relaxation times is necessary to describe the observed shear relaxation. The distribution of relaxation times is described by a asymmetric peak with a long tail extending to short times and a sharp cut-off at long times (Fig. 3.4). This is similar to the relaxation time spectra observed by other authors for silicate melts at high viscosities (i.e. Kurkjian, 1963; Mills, 1974). Each study of the relaxation of shear deformation in high viscosity silicate melts employs a different equation to describe the relaxation time spectrum. It is therefore very difficult to compare the fine details of the relaxation spectra between studies. In all cases however, the resulting spectrum can be described as "an asymmetric peak with a long tail extending to short times and a sharp cut-off at long times". Further comparison of the resulting spectra shows that the long tail extending to short times extends ~6-7 orders of magnitude away from the calculated Maxwell relaxation time and the sharp cut-off at long times occurs 2 orders of magnitude away from the Maxwell relaxation time. This distribution in relaxation times contrasts with the single relaxation time behaviour observed in silicate melts at lower viscosities using ultrasonic techniques (see Sect. 3.4).

This determination of the strain-rate dependent deformation of a rhyolite composition melt at high viscosities and small strains ($<10^{-5}$) shows that the onset of non-Newtonian rheology occurs 2 orders of magnitude slower than the calculated Maxwell relaxation time. This is in agreement with observations of the onset of non-Newtonian viscosity from ultrasonic studies but is not in agreement with Webb and Dingwell (1990a,b) in which the onset was found to occur 3 orders of magnitude slower than the Maxwell relaxation time. This suggests that the strains involved in fibre elongation determination of non-Newtonian rheology are indeed too large for the application of linear viscoelastic theory (Webb, 1992b).

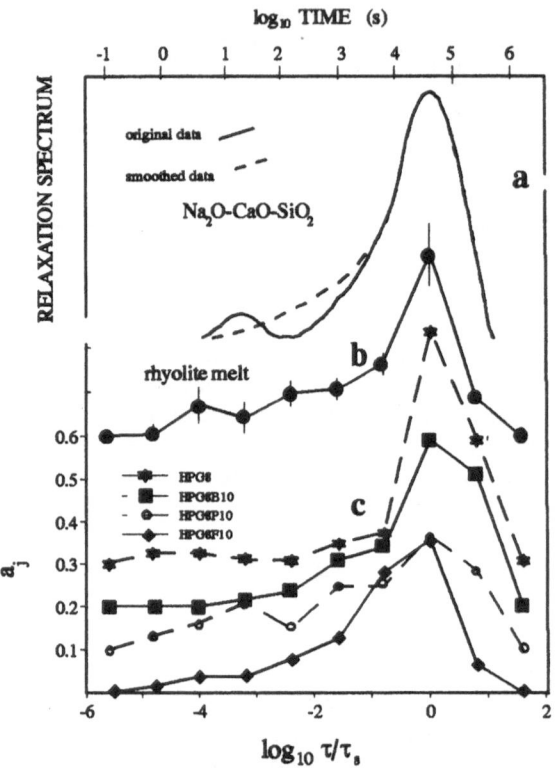

Fig. 3.4. a The relaxation spectrum for a Na_2O-CaO-SiO_2 composition melt - data from Kurkjian (1963). **b** The relaxation spectrum for rhyolite melt calculated from Eqn. 3.4 - data from Webb (1992a). **c** The relaxation spectra for granitic composition melts - data from Bagdassarov et al. (1993). [The spectrum for rhyolite melt has been shifted vertically by 0.6 units. Each of the curves for the granitic composition melts have been shifted vertically by 0.1 units.]

3.3.1 Composition Dependence of Shear Relaxation in Granitic Melts

The structure of silicate melts can be investigated using this forced torsion method as the full relaxation spectrum from relaxed liquid rheology to unrelaxed glassy rheology can be determined with some accuracy. The relaxation spectra for a series of granitic composition melts with small additions of phosphorus, boron and fluorine to the original melt composition have been investigated by Bagdassarov et al. (1993). The observed changes in the relaxation time spectrum in the glass transition region indicate the changes occurring in the Si-O structure as a function of the changing composition.

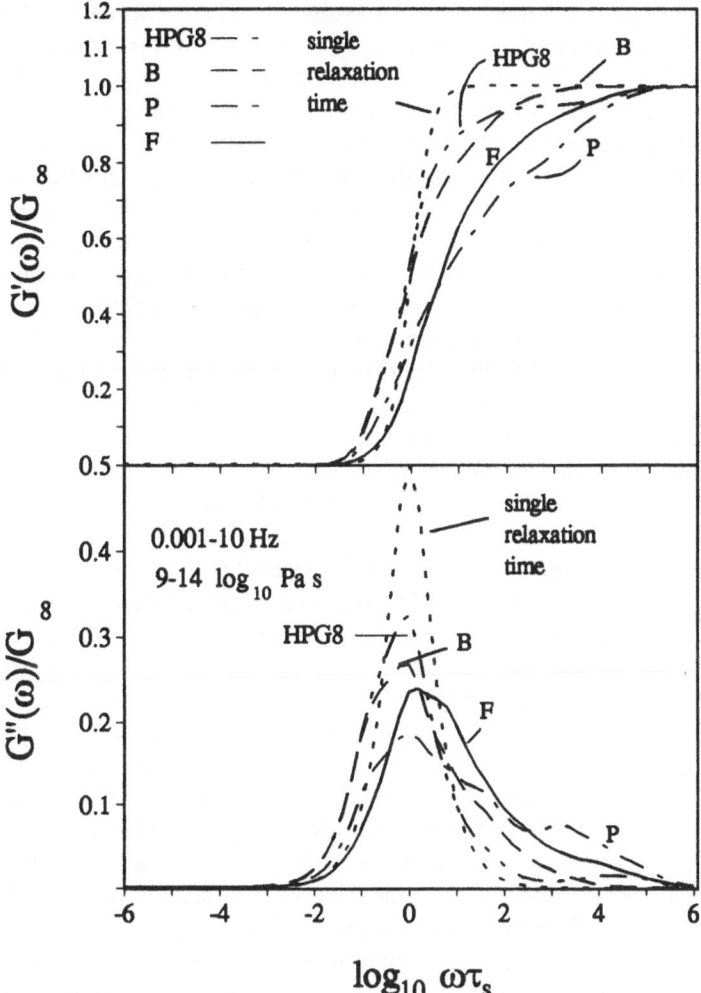

Fig. 3.5. The normalised complex real and imaginary shear moduli for HPG8 melt together with P$_2$O$_5$-, F$_2$O$_{-1}$, and B$_2$O$_3$-bearing HPG8 melt compositions - data from Bagdassarov et al. (1993).

Several components of silicate melts that are enriched in granitic magmas are known to have strong effects on the transport properties (i.e. viscosity, diffusion) of the melt. Bagdassarov et al. (1993) have investigated the effect on the shear modulus, shear viscosity and distribution of relaxation times for shear deformation of additions of P_2O_5, B_2O_3 and F_2O_{-1} to a granitic melt composition (HPG8: 84.7 mol% SiO_2, 4.8 mol% Na_2O, 2.9 mol% K_2O, 7.6 mol% Al_2O_3) using the forced oscillation technique. The addition of these oxides to the original HPG8 melt resulted in a decrease of up to 15% in the high temperature unrelaxed shear modulus G_∞. The unrelaxed bulk (volume) and shear modulus of the glasses at room temperature determined using ultrasonic interferometry were also decreased by up to 5% and 13 % respectively. The relaxed Newtonian shear viscosities of the melts decrease with the addition of these oxides by up to ~3 $\log_{10}$ Pa s at a temperature of ~1073 K (see Fig. 3.6).

The relaxation time spectra for the phosphorus-, boron-, and fluorine-bearing melts was shifted to lower temperatures as the inevitable consequence of the decrease in shear viscosity. More importantly, the relaxation spectra is seen to vary as a function of composition (see Fig. 3.4). With respect to the relaxation spectrum for the HPG8 melt, the phosphorus-, boron- and fluorine-bearing melts have relaxation time distributions which are skewed towards faster relaxation times (see Fig. 3.4). This broader distribution of relaxation times indicates an increase in the number of Si-O bond environments or bond strengths associated with the incorporation of the boron, phosphorus and fluorine ions in the structure of the melt. Again the Maxwell relationship can be used to estimate the relaxation time of the melt structure. There is no observable change in the long timescale relaxation spectrum, although the structure of the melt has been drastically altered by the addition of these components. This means that the onset of non-Newtonian behaviour as a function of increasing strain-rate will still be observed two orders of magnitude away from the relaxation strain-rate.

The addition of these components to the granitic melt composition, weakens the structure of the melt, making it less rigid (the shear modulus decreases). This is consistent with the skewing of the relaxation spectra to faster times, in that there exist a range of weaker Si-O bonds with shorter lifetimes as a result of the addition of boron, phosphorus and fluorine to the melt. The interpretation of the data in the forced oscillation measurements also relies on the use of thermorheological simplicity. The smooth lines which result from the application of thermorheological simplicity to the data indicate that this is indeed applicable in these small (120 K) temperature ranges (see Fig. 3.5.)

Once an understanding of the composition dependence of the rheology and relaxation of silicate melts exists we can extend our research towards more complex systems, specifically the effect of crystal and bubbles on the rheology of silicate melts. Bagdassarov et al. (1994) have investigated the strain-rate dependent rheology of synthetic crystal-bearing melts using the forced torsion method. Small spherical single crystals of Al_2O_3 were added to a melt of rhyolitic composition. It was not possible to synthesise these crystal-bearing melts without the coexistence of bubbles. The overall effect of these crystals and bubbles was to decrease the measured Newtonian shear viscosity of the sample. This forced oscillation technique is suited to determining the viscosity of bubble- and crystal-bearing melts as the strains involved are very small and the bubbles are not deformed. It is to be expected however, that with larger strains (i.e. the 1000% strains occurring in natural lava flows) the bubble-bearing melts will exhibit different rheology to that determined at much lower strains. It was not possible to observe a frequency independent viscosity in the one bubble-free sample with more than 45 vol% crystal content, indicating that in this range of crystal content the interaction between the crystals does not allow Newtonian flow of the melt even at very low deformation rates.

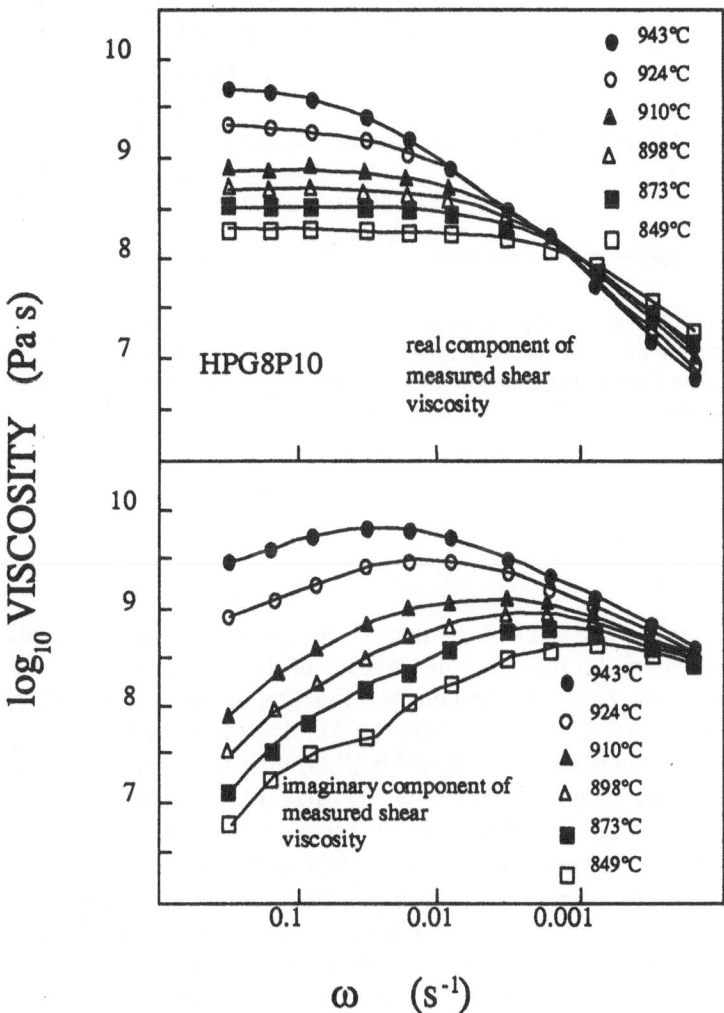

Fig. 3.6. The real and imaginary frequency-dependent shear viscosity for HPG8 + P_2O_5. (redrawn after Bagdassarov et al., 1993)

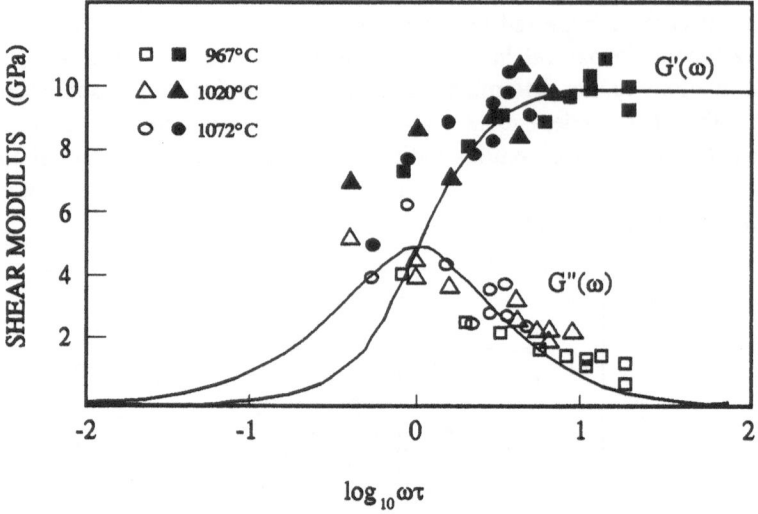

Fig. 3.7. The frequency-dependent shear modulus $G^*(\omega) = G'(\omega) + i\,G''(\omega)$ for $Na_2Si_2O_5$ melt in the viscosity range 1.3 - 2.3 $\log_{10}$ Pa s as a function of normalised angular frequency. The shear relaxation time for each temperature is calculated from the Maxwell relationship with $G_\infty = 10$ GPa. (redrawn after Webb, 1991)

3.4 Ultrasonics

Ultrasonic measurements involve strains of 10^{-8} and stresses of ~kPa and result in the observation of the frequency dependence of the velocity of a wave traveling through the melt. The shear modulus is a function of the melt density ρ and the wave velocity v; $G_\infty = \rho v^2$. Webb (1991) has determined the frequency dependence of the shear modulus of $Na_2Si_2O_5$ melt in the viscosity range $10^{1.3}$ - $10^{2.3}$ Pa s and the frequency range 5 to 150 MHz ($\omega = 30$ - 940×10^6 rad s^{-1}). The shear viscosity determined by this ultrasonic technique is identical to that determined by conventional concentric cylinder techniques for a melt of this composition. The measured frequency-dependent shear modulus and shear viscosity relaxation be described by Eqns. 3.2 and 3.3 for a single relaxation time process (see Figs. 3.7 and 3.8). The relaxation spectrum observed here is identical to that of a single relaxation time and suggests that the relaxation time spectrum at low viscosities is that of a single time and with increasing viscosity, the relaxation spectrum expands to include shorter (faster) relaxation times. As seen in Fig. 2.5, non-Newtonian viscosity has been observed in $Na_2Si_2O_5$ melt over a range of timescales using fibre elongation, torsion and ultrasonic methods. The smallest stress required to produce non-Newtonian viscosities in the fibre elongation measurements is 10^8 Pa. At higher temperatures (and lower viscosities) non-Newtonian viscosities are observed at much lower stresses; i.e. 10^5 Pa in torsion measurements and 10^3 Pa in ultrasonic measurements. This observation of non-Newtonian rheology occurring for lower stresses with decreasing viscosity is the opposite to the viscosity/stress trend that is predicted from the Adam-Gibbs model for non-Newtonian viscosity as discussed by Bottinga (1994) and Richet and Bottinga (1995). (The Adam-Gibbs model assumes that non-Newtonian viscosity is stress-related rather than rate-related.)

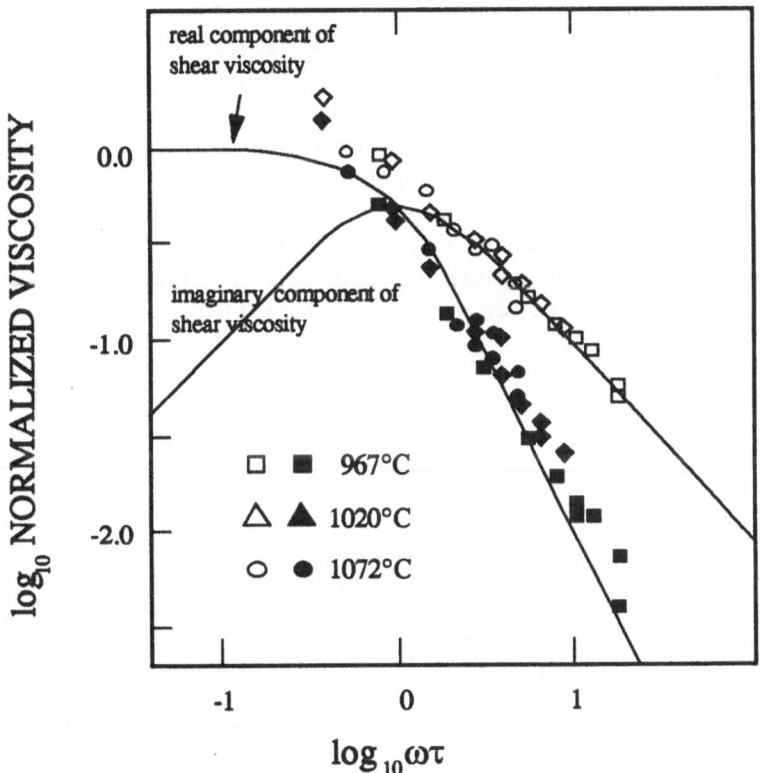

Fig. 3.8. The frequency-dependent shear viscosity $\eta_s(\omega) = \eta_s'(\omega) + i\eta_s''(\omega)$ for $Na_2Si_2O_5$ melt as a function of normalised angular frequency. The shear viscosity is also normalised to the Newtonian viscosity of the melt. The curve is calculated from the Eqn. 3.2. The $\log_{10}$ of the absolute value of the imaginary viscosity term is plotted. (redrawn after Webb, 1991)

The onset on non-Newtonian rheology at timescales two orders of magnitude slower than the Maxwell relaxation time is in agreement with observation made on the frequency dependent rheology of silicate melts observed at high viscosities by forced oscillation techniques (Webb, 1992b; Mills, 1974; Kurkjian, 1963). The forced oscillation data of Mills (1974) on a melt of the same composition, however, observed at high viscosities that the relaxation time distribution had a long tail extending to times faster than the Maxwell relaxation time. The single relaxation time data of Webb (1991) for $Na_2Si_2O_5$ melt at shear viscosities of $10^{1.3}$ to $10^{2.3}$ Pa s in combination with the relaxation time data of Mills (1974) would indicate that the extent to which the viscoelastic regime extends to timescales faster than the calculated Maxwell relaxation time depends upon the viscosity of the melt. The regime in which relaxed, unrelaxed and viscoelastic behaviour should be expected for silicate melts in general is sketched as a function of viscosity and relaxation time in Fig. 3.9. This suggests that for high viscosity melts, there exists a range of relaxation times which collapse to one at lower viscosities. This points to either a range of Si-O bonds or bond strengths or activation energies of Si-O bond exchange at low temperature/high viscosity conditions which are no longer distinguishable at low viscosities (high temperatures). The onset on non-Newtonian rheology in geological processes as a function of strain-rate will be uninfluenced by the absolute value of shear viscosity. The effects of structural relaxation in cooling melts will extend over large ranges of temperature below the glass transition temperature into the glassy region of melt behaviour.

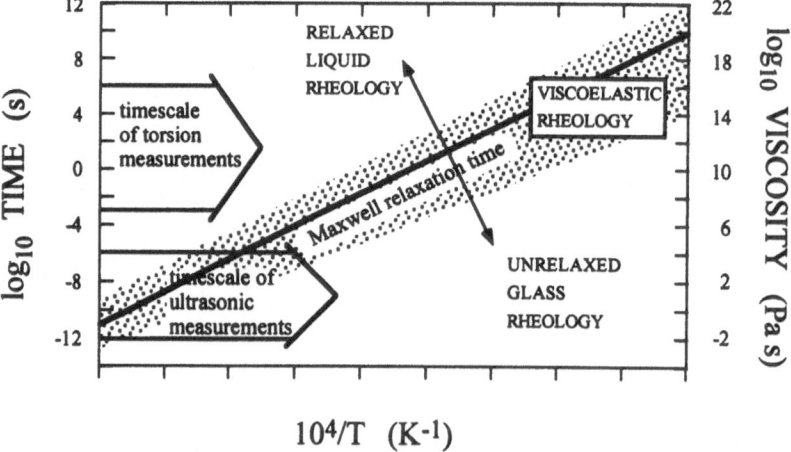

Fig. 3.9. The region of viscoelastic rheology in temperature-time space based on the data of Kurkjian (1963), Mills (1974), Rivers and Carmichael (1985) and Webb (1991) obtained for a range of silicate melt compositions.

3.5 Shear Viscosity

In general, shear viscosities of silicate melts are determined by concentric cylinder techniques at low viscosities in the range 10^0 - 10^5 Pa s. There are a range of techniques which allow the determination of very high viscosities in the range 10^8-10^{14} Pa s. The data obtained in this viscosity range have always been plagued by the problem of structural relaxation of the melt, which results in partially unrelaxed (i.e. lower) viscosities being determined as illustrated by Lillie (1933), Rekhson (1989) and Richet and Bottinga (1995). As shown above (and in the following) viscosity is a measure of the timescale on which the Si-O structure of a melt relaxes. Therefore the temperature and composition dependence of shear viscosity is an important indicator for changes in melt structure, and changes in other physical properties of silicate melts. At low viscosity conditions it has been observed that the viscosity is not very sensitive to changes in composition, but at high viscosity conditions, the observed viscosity is very sensitive to small changes in composition. A wide range of geological processes occur at very high viscosities (e.g. crystal-melt fractionation in granitic composition melts at $\eta \geq 10^9$ Pa s at ~850 K). Therefore it is necessary to accurately determine viscosities greater than 10^8 Pa s. This can be achieved by ensuring that the timescale of measurement is at least two orders of magnitude longer than the structural relaxation time of the melt. For melts of viscosity 10^9 Pa s, $\tau \sim 0.1$ s (implying that the structure is fully relaxed for times greater than 10 s) and measurement times should therefore be greater than 10 s duration.

Fibre elongation techniques are suitable to determine these high viscosities (Webb and Dingwell, 1990a,b) but the sample preparation involves remelting the ends of the melt fibre to fit into the sample holder. This process results in the loss of volatiles from the melt or in some cases the crystallisation of the fibre. Some fibre elongation geometries involve long fibres (~100 mm) in order to overcome the possible effects of composition changes at the ends of the fibres. These types of measurement however, have the problem of a thermal gradient across the length of the fibre. For the case of F_2O_{-1} it was impossible to measure viscosity in a concentric cylinder viscometer due to loss of fluorine (Dingwell et al., 1993b). A technique which requires no further low viscosity heating is micropenetration. This involves forcing an indenter into a semi-polished slice of the melt in the viscosity range 10^9 - $10^{11.5}$ Pa s. This technique has been used to further study the high viscosity of HPG8 with additions of various oxides (Dingwell et al., 1993b,c; Hess et al., 1995a,b). The relaxed shear viscosities determined for these HPG8-X melts can then be compared. Figure 3.10 illustrates the agreement between the shear viscosity of HPG8 + B_2O_3 determined using micropenetration and torsion techniques in the range 10^9 - $10^{11.5}$ Pa s. The inclusion of concentric cylinder data illustrates the Arrhenian behaviour of the viscosity from $10^{10.5}$ - 10^2 Pa s.

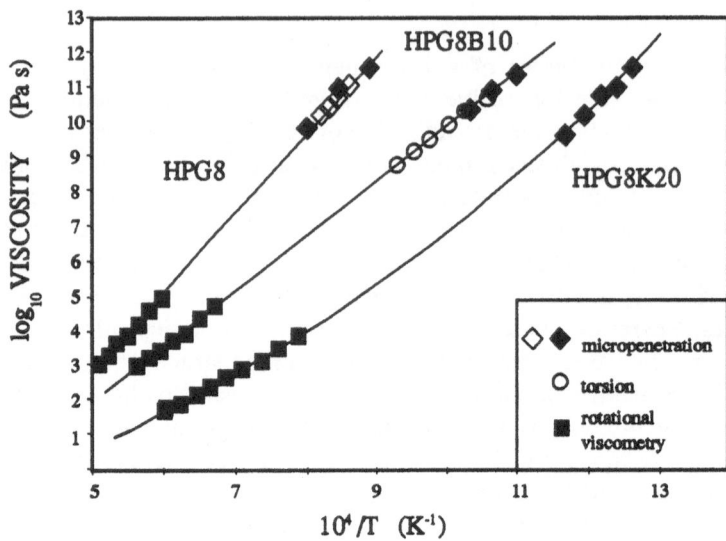

Fig. 3.10. Relaxed shear viscosity as a function of inverse temperature determined using concentric cylinder, micropenetration and forced oscillation techniques. HPG8 data from Dingwell et al. (1992) - solid symbols; and Hess et al. (1995a) - hollow symbols; HPG810B (19 wt% added B_2O_3) data from Bagdassarov et al. (1993) and HPG8K20 (20 wt% added K_2O) data from Hess et al. (1995a).

The parallel plate deformation technique also requires no heating of the sample at low viscosity conditions. This involves applying a stress to a cylinder (or parallelopiped) of melt and determining the resulting rate of deformation of the melt. Bagdassarov and Dingwell (1992), Lejeune and Richet (1995), Stevenson et al. (1995) and Stevenson et al. (1996) have used this method to measure viscosities of natural and synthetic silicate melts in the range 8 - 12 $\log_{10}$ Pa s. Viscosity is determined by deformation of large samples (e.g. 20 mm long, 8 mm diameter cylinders) by < 5% along the long axis. This technique can be used to investigate the viscosity of both crystal- and bubble-bearing melts at low strain conditions. Spera et al. (1988) and Stein and Spera (1992) have demonstrated that the viscosity of bubble-bearing melts can also be determined at high-strain (e.g. 1000%) conditions using a concentric-cylinder technique in the viscosity range 10^4 - $10^{6.5}$ Pa s.

4 Volume Relaxation

The timescale of volume relaxation can be determined by a variety of techniques either at a constant temperature as a function of observation time or as a function of heating-rate across the glass transition. Dilatometric techniques to determine the volume change as a function of heating-rate result in a volume versus temperature curve which has the shape of Fig. 2.1a. The dV/dT curve follows the curve of Fig. 2.1b. The relaxation of volume can be determined at constant temperatures by ultrasonic measurements of the frequency dependence of the longitudinal and bulk moduli. Early studies in which the longitudinal modulus was determined illustrate that the relaxation time for volume relaxation is identical to that of shear relaxation. Ultrasonic studies in which both shear and longitudinal moduli of silicate melts were determined, allowed the calculation of the bulk modulus as a function of frequency, with the result that the relaxation times for shear and volume are found to be the same, within error of the measurements, at these temperatures. This type of measurement also gives information about volume viscosity. In contrast to shear viscosity which can occur for an infinite strain (without change in volume) volume viscosity must occur over a limited strain. Logically it is expected that anelastic (as opposed to viscoelastic) volume deformation occurs in silicate melts. The imaginary part of the volume modulus is then treated mathematically as a viscosity term.

4.1 Anelasticity

In the case of volumetric deformation, silicate melts behave as anelastic materials. Only instantaneous and delayed recoverable deformation are observed, with no time dependent non-recoverable deformation occurring. Both the instantaneous and long timescale observation of melt deformation result in the determination of Newtonian rheology (strain-rate independent deformation). However, if the deformation of the melt is observed on a timescale similar to that on which the delayed recoverable deformation takes place non-Newtonian time-dependent (strain-rate dependent) linear viscoelastic rheology of the melt will be observed.

The change in melt behaviour from elastic to anelastic is due to the relaxation of the melt structure. In the cases where the volume relaxation time has been determined, the frequency-dependent volume (bulk) modulus can be described by;

$$K^*(\omega) = K_0 + \frac{K_1\omega^2\tau_v^2}{1+\omega^2\tau_v^2} + i\,\frac{K_1\omega\tau_v}{1+\omega^2\tau_v^2} \qquad (4.1)$$

(Herzfeld and Litovitz, 1959) where the elastic (infinite frequency) modulus $K_\infty = K_0 + K_1$; and the Maxwell relationship defines the relaxation time for volume deformation

$$\tau_v = \frac{\eta_v}{K_1} \qquad (4.2)$$

for the "volume viscosity" η_v and the relaxation component of the bulk modulus K_1 of the melt. The volume viscosity can be determined from the frequency dependent volume modulus (Webb, 1991);

$$\eta_v^* = \frac{K^*(\omega)}{i\omega} = \frac{K_1\tau_v}{1+\omega^2\tau_v^2} - i\left(\frac{K_0}{\omega} + \frac{K_1\omega\tau_v^2}{1+\omega^2\tau_v^2}\right) = \eta_v' + i\eta_v''. \qquad (4.3)$$

The volume viscosity is strain limited (Mazurin, 1986) but it accounts for ~1% increase in melt density per GPa pressure increase. It has been shown that for silicate melts in general that the approximation $\eta_v \approx \eta_s$ holds (Dingwell and Webb, 1990).

4.2 Ultrasonics

The compressibility of silicate melts is a subject of great importance to the understanding of magma ascent and eruption within the Earth and terrestrial planets. Accordingly several experimental studies of the elastic properties of silicate melts have been carried out over the past decade. The relaxed bulk modulus of silicate melts has been studied for a wide variety of silicate melt compositions and calculational models have been derived for the estimation of multicomponent melt compressibilities (Rivers and Carmichael, 1987; Kress and Carmichael, 1991; Webb and Courtial, 1996). Most ultrasonic studies of silicate melts are conducted at the high temperature (1200-1800 K) conditions and low frequencies (3-22) MHz required to observe the relaxed (frequency-independent) longitudinal (= volume) modulus of the melt. In the cases where the experimental conditions approached the relaxation frequency of the melt (e.g. Manghnani et al., 1981; Rivers and Carmichael, 1987; Sato and Manghnani 1985; Kress et al., 1988; Secco et al., 1991; Bornhöft and Brückner, 1994) frequency-dependence of the longitudinal modulus has been observed. In studies where the frequency of the applied signal has been greater than that of the relaxation frequency, shear wave propagation has been observed [Macedo et al., 1968 (Na_2O-B_2O_3-SiO_2 melt); Webb, 1991 ($Na_2Si_2O_5$ melt)].

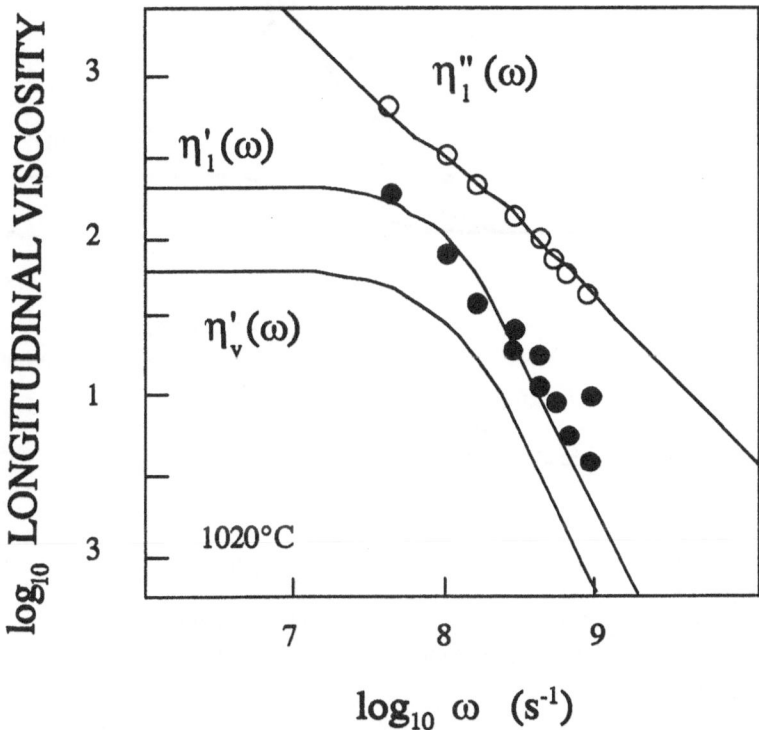

Fig. 4.1. The frequency-dependent longitudinal and volume viscosity of $Na_2Si_2O_5$ melt as a function of frequency. The log_{10} of the absolute value of the imaginary viscosity term is plotted. There is no appreciable difference between $\eta_l''(\omega)$ and $\eta_v''(\omega)$. The curves are calculated from Eqn. 4.4. (redrawn after Webb, 1991)

In the study of Webb (1991) the viscosity of the melt (10^1- 10^3 Pa s) and the frequency of the ultrasonic signal (5-150 MHz) were chosen such that both the relaxed and unrelaxed longitudinal (M) and shear (G) moduli and viscosities of $Na_2Si_2O_5$ could be measured with the volume viscosity and modulus being determined from Eqn. 4.4.

$$\eta_l^*(\omega) = \eta_v^*(\omega) + 4\eta_s^*(\omega)/3. \qquad\qquad (4.4)$$

$$M^*(\omega) = K^*(\omega) + 4G^*(\omega)/3. \qquad\qquad (4.5)$$

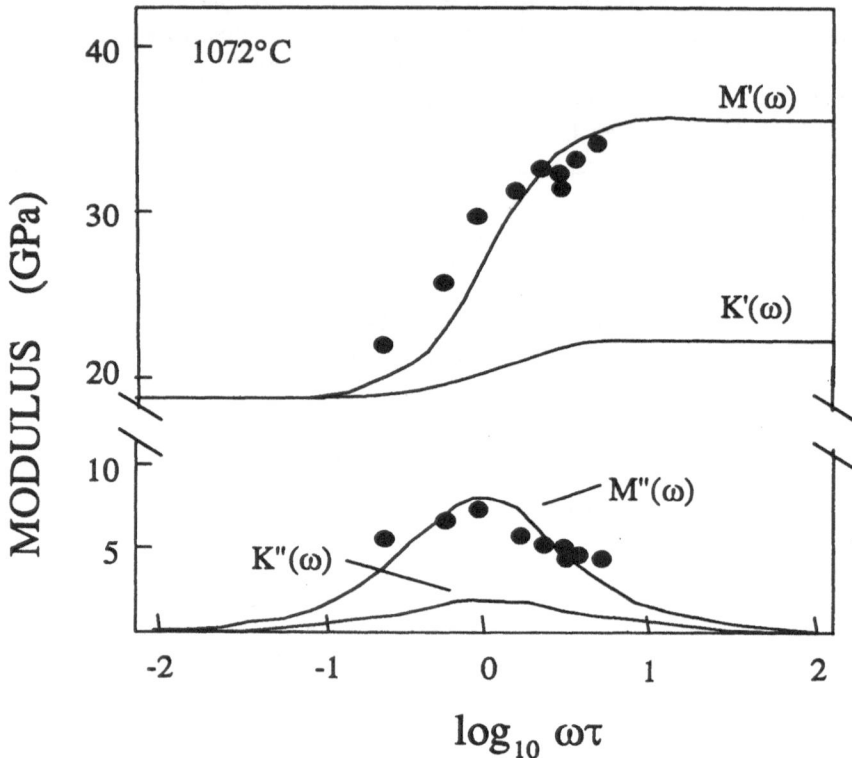

Fig. 4.2. The frequency-dependent longitudinal and volume modulus of $Na_2Si_2O_5$ melt as a function of normalised frequency. The curves are calculated from Eqn. 4.4. (redrawn after Webb, 1991)

The frequency-dependent volume modulus and viscosity of $Na_2Si_2O_5$ melt could them be determined from the measured shear and longitudinal moduli and viscosities of $Na_2Si_2O_5$ melt. The measured longitudinal viscosity and calculated volume viscosity of $Na_2Si_2O_5$ melt are illustrated in Fig. 4.1. The relaxed volume modulus is 20 GPa with the unrelaxed modulus being 25 GPa (see Fig. 4.2). No distribution in relaxation time could be resolved from the frequency dependent modulus and viscosity data, with all indications being that the data could be fit with a single relaxation time curve for the shear and longitudinal data and the calculated volume data. Bornhöft and Brückner (1994) have also studied the viscoelastic rheology of four silicate melts using ultrasonic techniques in the frequency range 0.5 to 4 MHz and the viscosity range 10^1-10^5 Pa s. Askarpour et al. (1993) have demonstrated that the relaxed and unrelaxed volume and shear moduli of silicate melts can be determined at a fixed frequency using Brillouin scattering techniques.

4.2.1 Titanium-bearing silicate melts

The composition dependent structure of melts can be investigated as a function of anomalous increase or decrease in relaxed bulk modulus. An example is the bulk modulus of titanosilicate melts. Titanium oxide appears as a component of minor to major element concentration in the chemistry of the vast majority of terrestrial and extraterrestrial magmatic rocks. Of all the major elements occurring in terrestrial lavas, titanium has the distinction of having a strongly compositionally dependent structural role in silicate melts. This variation in co-ordination has measurable consequences for melt properties. The presence of Ti in alkali-titanosilicate melts results in anomalous negative temperature dependence of heat capacity in the 400 K above the glass transition temperature (Richet and Bottinga, 1985) and a doubling of the Δc_p at the glass transition (Lange and Navrotsky, 1993). The large step in heat capacity at T_g reflects the significant configurational rearrangements in the melt which do not occur in Ti-free melts. The decrease in heat capacity with temperature indicates that these configurational changes cease at higher temperatures. The linearly extrapolated partial molar volume of TiO_2 in alkali-titanosilicate melts varies by 15% as a function of the alkali identity (Dingwell, 1992). The partial molar volume of TiO_2 calculated for the alkaline-earth titanosilicate melts is independent of cation identity. The low partial molar volume of TiO_2 in these melts has been taken to imply that Ti is mainly in octahedral co-ordination (Dingwell, 1992). In view of this anomalous heat capacity and composition dependent partial molar volume of TiO_2 observed in the alkali-titanosilicate melts, it is expected that the co-ordination state of titanium will affect the physical properties (e.g. density, thermal expansion, heat capacity, compressibility) of titanium-bearing melts in a non-linear manner as a function of both temperature and composition.

The effect of composition on the relaxed adiabatic bulk modulus (K_0) of a range of alkali- and alkaline earth-titanosilicate [$X_{2/n}^{n+} TiSiO_5$ (X = Li, Na, K, Rb, Cs, Ca, Sr, Ba)] melts has been investigated (Webb and Dingwell, 1994). These melts have very low viscosities [e.g. 0.02-5 Pa s] in the temperature range 1100-1900 K, such that the relaxed bulk moduli of the melts can be determined using ultrasonic techniques in the 3-7 MHz range. For these frequencies and viscosities, the period of the sinusoidal acoustic wave is always at least 2 orders of magnitude more than the structural relaxation time of the melt and therefore the relaxed velocity of a compressional wave propagating through the melt is measured. The bulk moduli of these melts decrease with increasing cation size from Li to Cs and Ca to Ba; and with increasing temperature.

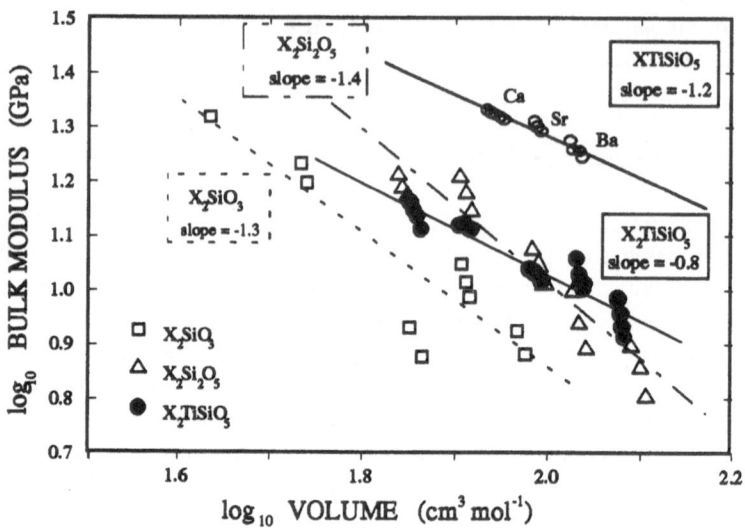

Fig. 4.3. $KV^{4/3}$ relationship for X_2SiO_3, $X_2Si_2O_5$ and X_2TiSiO_5 (X = Li, Na, K, Rb, Cs) and $YTiSiO_5$ (Y = Ca, Sr, Ba) melts. (Data from Webb and Dingwell, 1994)

Linear extrapolation to TiO$_2$. Selecting data for the metasilicate and titanosilicate melt compositions at temperatures of 1273 K (alkali melts) and 1873 K (alkaline-earth melts) one can attempt to linearly extrapolate as a function of volume% TiO$_2$ in the melt to the bulk modulus of the TiO$_2$ component. It has been shown by Watt et al. (1976) for crystalline materials that the bulk modulus of the mixture is bounded by the sum of the bulk moduli (K_i) as a function of volume fraction (v_i) and the sum of the compressibilities (K_i^{-1}) as a function of volume fraction

$$\left(\sum_{i=1}^{n} v_i / K_i \right)^{-1} \leq K \leq \sum_{i=1}^{n} v_i K_i . \qquad (4.5)$$

The bulk modulus for TiO$_2$ is found to range from 0.7-40 GPa at 1273 K to 3-15 GPa at 1873 K, calculated from Eqn. 4.5. These linear fits to the bulk modulus versus volume fraction of TiO$_2$ do not converge to a common compressibility of the TiO$_2$ component. This indicates that the structural role of TiO$_2$ in these melts is dependent on the identity of the cation. The composition dependence of the compressibility of the TiO$_2$ component in these melts explains the difficulty incurred in previous attempts to incorporate TiO_2 in calculation schemes for melt compressibility (see Kress and Carmichael, 1991). Such a variation in the structural role and compressibility of the titanium component in silicate melts makes it impossible to calculate the bulk modulus of TiO_2 liquid.

Iso-structural Melts. In order to investigate the structural changes occurring in these melts as a function of composition the bulk modulus can be plotted as a function of the molar volume of the melt. A semi-empirical relationship between the bulk modulus, K, and the molar volume, V, (or the volume per ion pair, V_{pair}) for iso-structural crystalline materials has been developed from classical ionic physics;

$$K \propto V^{-4/3} \tag{4.6}$$

(Bridgman, 1923; Anderson and Nafe, 1965). This relationship has been found to apply in general to iso-structural oxides and silicates and Rivers and Carmichael (1987) and Herzberg (1987) have applied this relationship in discussions of the compressibility of silicate melts. Both of these studies pointed out that this relationship is applicable to simple binary alkali-silicate melts but not to alkaline-earth compositions.

Figure 4.3 is a log/log plot of the relaxed bulk modulus versus molar volume for alkali-metasilicate, alkali-disilicate and alkali- and alkaline-earth-titanosilicate composition melts. The slope of the straight lines fitted to the alkali-metasilicate and alkali-disilicate melt compositions is -4/3 implying that the melts in each of these series are iso-structural. The slope of the straight line fitted to the alkali-titanosilicate data is -0.8±0.1, implying that the structure of the melts changes as a function of composition. The structural parameter causing the shift in melt modulus/volume systematics is likely to be the shift in average co-ordination number of Ti inferred from X-ray absorption studies of glasses quenched from these melts (Paris et al., 1993). The low slope points to there being an increasing amount of the compressible octahedrally co-ordinated titanium in the melts with the smaller alkali cations, with respect to the less compressible tetrahedrally co-ordinated titanium. The slope of the straight line fitted to the alkaline earth-titanosilicate data is -4/3, implying that these titanium-bearing melts are iso-structural independent of cation identity with the co-ordination of Ti being a constant, independent of the alkaline-earth cation.

4.2.2 $CaO-Al_2O_3-SiO_2$ Melts

The majority of silicate melt compositions have a bulk modulus in the range 5 to 20 GPa at one atmosphere pressure. The alkaline-earth bearing melts have much higher bulk moduli, from 20-30 GPa. That is, they are not as compressible as the

alkali-rich silicate melts studied to date. As shown by Rivers and Carmichael
(1987) and Herzberg (1987) and illustrated in Fig. 4.4, the bulk moduli of CaO-
bearing silicate melts (and alkaline earth-bearing melts in general) do not follow
the simple empirical trends observed as a function of molar volume for most
silicate melts and needs to be investigated in more detail. The effect of
composition and temperature on the compressibility of melts in the CaO-Al_2O_3-
SiO_2 system has been investigated by Webb and Courtial (1996). For these melts,
the bulk modulus increases with the addition of CaO, and decreases with addition
of SiO_2 and Al_2O_3. The bulk moduli of these alkaline-earth aluminosilicate melts
are ~50% larger than the adiabatic bulk moduli of the sodium-aluminosilicate
melts of Kress et al. (1988).

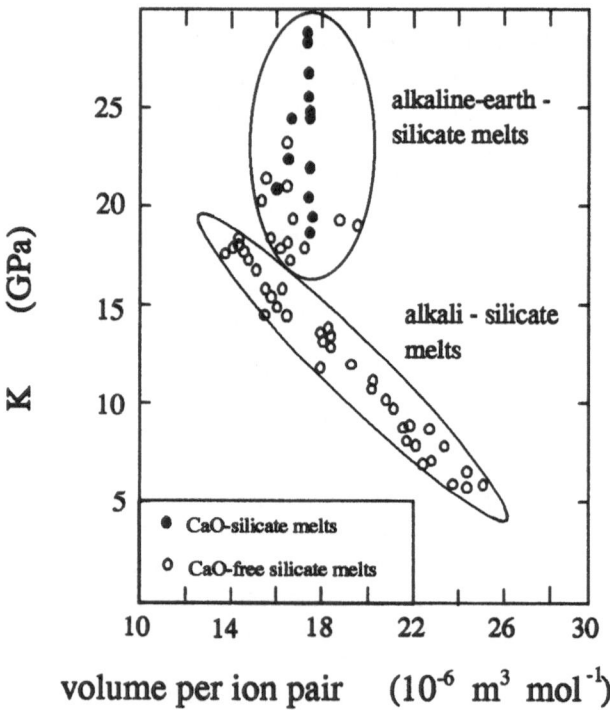

Fig. 4.4. Bulk modulus as a function of volume per ion pair for a range of silicate melt
compositions. The separate trends of the alkali- and alkaline earth-silicate melts are
indicated. (redrawn after Rivers and Carmichael, 1987)

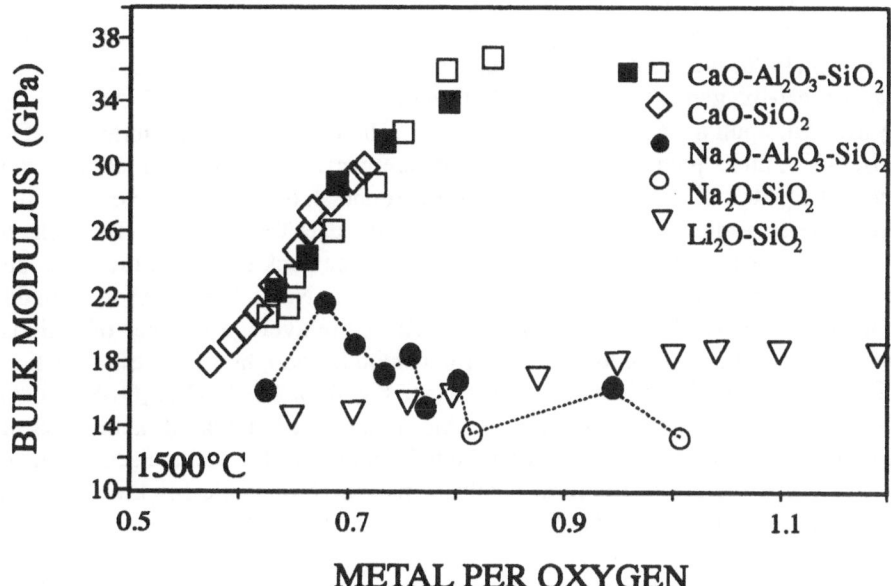

Fig. 4.5. Bulk modulus as a function of cation-per-oxygen for melts in the CaO-Al$_2$O$_3$-SiO$_2$, Na$_2$O-Al$_2$O$_3$-SiO$_2$ and Li$_2$O-SiO$_2$ systems. (redrawn after Webb and Courtial, 1996)

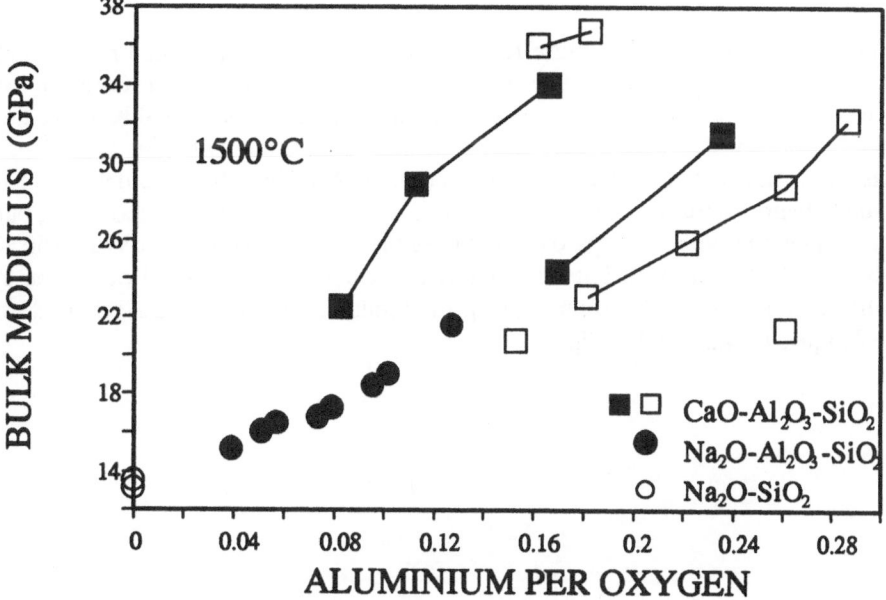

Fig. 4.6. Bulk modulus as a function of the number of aluminiums-per-oxygen for melts in the CaO-Al$_2$O$_3$-SiO$_2$ and Na$_2$O-Al$_2$O$_3$-SiO$_2$ systems. Data of constant Ca:Al are joined by solid lines. (redrawn after Webb and Courtial, 1996)

In a further attempt to understand the relationship between structural geometry, volume and compressibility in the calcium-aluminosilicate melts, Webb and Courtial (1996) plotted the bulk modulus at 1723 K as a function of the number of cations-per-oxygen (Fig. 4.5). The bulk modulus data for the CaO-Al_2O_3-SiO_2 system fall upon a single smooth in Fig. 4.5. The behaviour of the ternary CaO-Al_2O_3-SiO_2 melts points to the variation in compressibility being a function of the average number of cations which need to fit around an oxygen. As the data for the Al_2O_3-bearing melts fall on the same curve as the data for the Al_2O_3-free melts it would appear that it is the CaO which controls the packing of the metal cation around the oxygens. The bulk moduli of melts in the Na_2O-Al_2O_3-SiO_2 system do not follow a smooth variation as a function of the average number of cations around an oxygen. The bulk moduli form a smooth trend, however, as a function of the number of aluminiums-per-oxygen (see Fig. 4.6). The CaO-Al_2O_3-SiO_2 data fall on separate curves of constant CaO:Al_2O_3 ratio as a function of the number of aluminiums-per-oxygen. The data on the bulk moduli of the melts discussed here indicate that different structural mechanisms are responsible for the observed compressibility of CaO-Al_2O_3-SiO_2 and Na_2O-Al_2O_3-SiO_2 melts.

4.3 Dilatometry

The glass transition temperature for volume can also be determined by thermal expansion measurements. The volume and dV/dT data as a function of temperature is identical to Fig. 2.1a,b. The expansion data some 5-10 K above the peak in the dV/dT curve are no longer reliable as the sample begins to collapse viscously under its own weight (Webb et al., 1992). Therefore the density of a melt at temperatures just above T_g cannot be determined by this method alone; but in conjunction with enthalpy data it is possible to calculate V and dV/dT ~50 K above T_g (see Sect. 5.2). The temperature range over which thermal expansion in silicate melts can be determined using traditional dilatometry and Archimedean techniques is illustrated in Fig. 4.7.

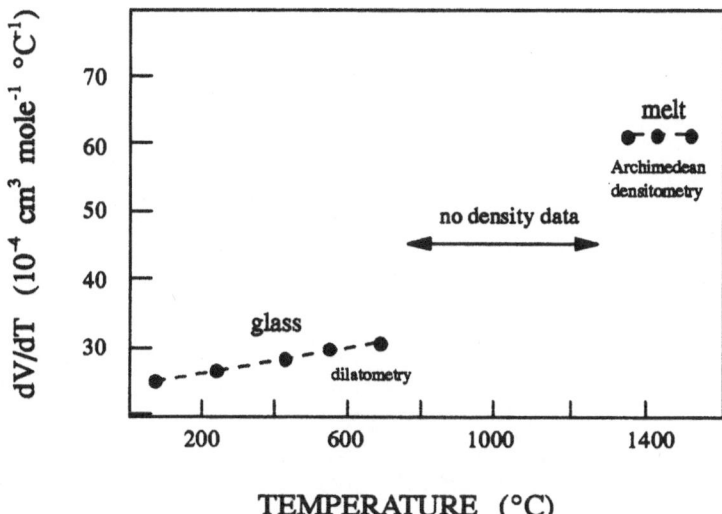

Fig. 4.7. Temperature range of thermal expansion determinations by dilatometry and high temperature densitometry.

5 Enthalpy Relaxation

Structural relaxation is discussed here in terms of variation in the macroscopic property enthalpy. Enthalpy relaxation can be observed in silicate melts as a function of temperature at a fixed timescale of observation, or as a function of observation timescale at a fixed temperature. At a fixed heating rate, the relaxation of configurational enthalpy can be observed as the structure of the melt equilibrates with the temperature of the melt (see Crichton and Moynihan, 1988; Narayanaswamy, 1971, 1988; Scherer, 1984, 1986). At a fixed temperature the linear response of the enthalpy of a system in equilibrium to a small temperature oscillation can be observed (e.g. Birge and Nagel, 1985). Due to the relative ease of measuring the heat capacity (Δ[enthalpy]/ΔT) of silicate melts via scanning calorimetry, the following discussion of enthalpy relaxation is based on data obtained as a function of cooling- and heating-rate.

The glass transition temperature T_g for enthalpy relaxation can be obtained from scanning calorimetry data on melts. The heat capacity curve ($c_p = dH/dT$) has the same shape as the general property curve and the dP/dT curve in Fig. 2.1a,b; see Fig. 5.1. There are a number of different definitions of T_g which are used in the scanning calorimetry literature. All of these definitions are based on the position of the peak in the c_p curve. T_g is often taken to be the extrapolated onset of the heat capacity peak, the inflection point in the rapidly rising part of the heat capacity curve or the maximum in the heat capacity curve (see Moynihan et al., 1974). The peak in the dH/dT curve is taken to be T_g in the present discussion, in order to be consistent with the definition of T_g used in the volume relaxation measurements presented in Sect. 4.3.

There has been some discussion in the literature about whether the T_g determined using different techniques, and the T_g obtained for different properties in silicate melts are identical. The studies of Moynihan et al. (1974; 1976a,b) indicate that within the errors of the temperature measurement, the T_gs for enthalpy, volume and shear relaxation are the same. Comparison of T_g data determined using different heating-rates and different definitions has led to some confusion about these observations of Moynihan et al., but as shown by Webb and Knoche (1996) the T_gs determined for a large number of silicate melts by both scanning calorimetry and dilatometry are the same.

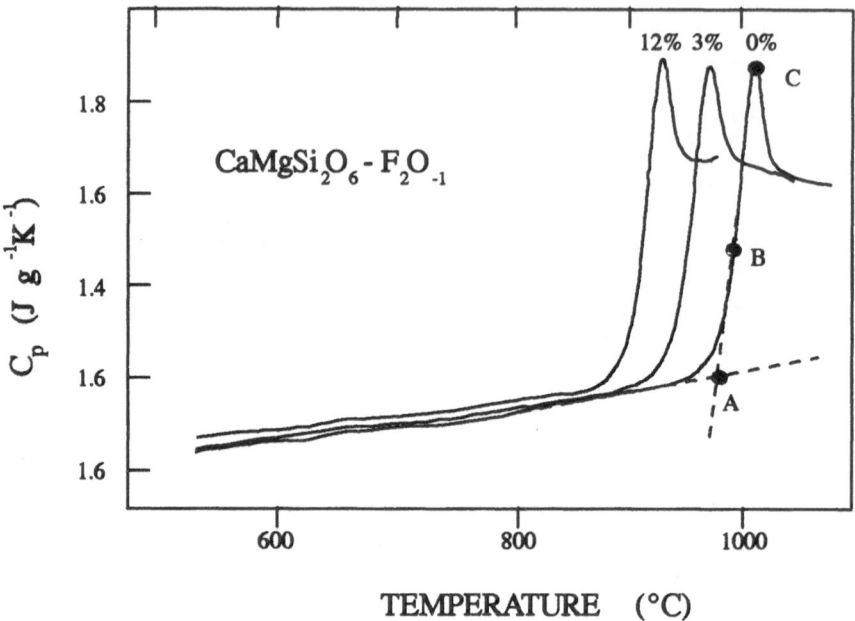

Fig. 5.1. Scanning calorimetric traces of the heat capacity of glassy and liquid melts on the join diopside-F_2O_{-1} cooled and subsequently reheated at 5 K min^{-1}. (Redrawn after Dingwell and Webb, 1992)

5.1 Enthalpy and Viscosity

It has been demonstrated for a number of silicate melts that calorimetric (dH/dT) and dilatometric (dV/dT) peak temperatures represent temperatures of constant viscosity or isokoms (Scherer, 1986) as a function of composition. This is because the temperature-dependent processes of volume and enthalpy relaxation are strongly linked to those controlling shear flow (Dingwell and Webb, 1989; 1990). The relaxation timescale for all of these processes can be estimated from the Maxwell relation (Dingwell and Webb, 1989) which relates the ratio of shear viscosity and shear modulus to the shear relaxation time. Literature data for the shear moduli of diopside and anorthite glasses (Bansal and Doremus, 1986) allow us to confirm this approximation for the diopside - anorthite system as illustrated in Fig. 5.2. The viscosity - temperature relationships of melts on the diopside - anorthite join are reproduced from Tauber and Arndt (1986). The calorimetric and dilatometric (volume) peak temperature for each composition (measured on samples that had been cooled at 5 K min^{-1} and heated at 5 K min^{-1}) has been plotted on the corresponding viscosity-temperature relationship. These temperatures are identical ±3 K which is the error in determining T_g from the calorimetry and dilatometry data. The resulting array of points describes an isokom with $\eta_s = 11.0\pm0.2$ log$_{10}$ Pa s. The isokom obtained in Fig. 5.2 leads to the conclusion that enthalpy, volume and shear relaxation vary sympathetically across the diopside - anorthite join and confirm the approximation of the Maxwell relation as a source of relaxation times. The glass transition, consistently measured and described in silicate melts, approximates an isokom (Knoche et al., 1992a). If cooling- and heating-rates other than 5 K min^{-1} are used, the viscosity at T_g will not be $10^{11.0}$ Pa s.

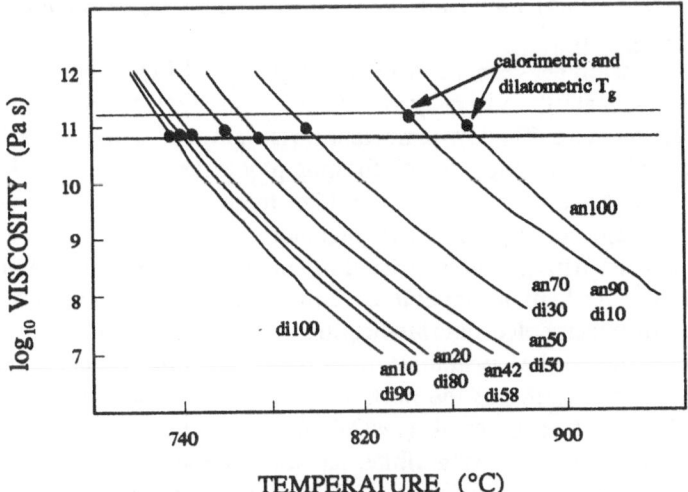

Fig. 5.2. Comparison of the peak temperature from calorimetry and viscosity along the diopside-anorthite join. The glass transition temperature T_g approximates an isokom. (redrawn after Knoche et al., 1992a)

Studies of enthalpy relaxation in silicate melts result in the conclusion that the activation energy required for enthalpy relaxation is identical with that of viscous flow (and volume relaxation). This leads to the following relationship between T_g and viscosity, that is independent of composition;

$$-\log_{10} |q| = A + \frac{B}{T_g}$$

and

$$\log_{10} \eta_s(\text{at } T_g) = C + \frac{B}{T} \tag{5.1}$$

(Moynihan et al., 1974). As the activation energy B is identical for both viscous flow and enthalpy relaxation (Rekhson et al., 1971; Moynihan et al., 1976a, b; Bornhöft and Brückner, 1994; Stevenson et al., 1995)

$$\log_{10} \eta_s(\text{at } T_g) = \text{constant} - \log_{10}|q| \tag{5.2}$$

where Scherer (1984) found the constant to be 11.3 for η_s in $\log_{10}$ Pa s and quench-rate $|q|$ in K s^{-1}.

This relationship between viscosity and T_g obtained from calorimetry data can be used to estimate the viscosity of silicate melts. One special case is that of volatile bearing melts. The low-temperature viscosities of fluorine-bearing silicate melts are of direct relevance to rock forming processes in granitic and pegmatitic environments. Attempts to extrapolate high temperature viscosity data to low temperatures relevant to granitic and pegmatitic petrogenesis are difficult as the high temperature data are too restricted in temperature range to account for any potential non-Arrhenian behaviour. Low temperature data are required. Viscosities of undercooled fluorine-bearing melts are difficult to measure at low temperatures due to the crystallisation of the melt (Dingwell and Webb, 1992). Based on the calibration of the peak temperature of melts along the diopside-anorthite join by Knoche et al. (1992a) the peak temperatures from scanning calorimetry of fluorine-bearing albite, jadeite and nepheline composition melts yield low temperature viscosity data. Thus the Arrhenian and non-Arrhenian nature of a variety of geological fluorine-bearing melts can be investigated (Dingwell and Webb, 1992).

Further investigation of this relationship between T_g obtained from scanning measurements and shear viscosity by Knoche et al. (1994) indicates that the viscosity at T_g for melts in the Na_2O-SiO_2 join from 55 to 80 mol% silica have a viscosity 10.8±0.2 $\log_{10}$ Pa s, in agreement with the 11.0±0.2 $\log_{10}$ Pa s observed in the study of Knoche et al. (1992a). For melts with more than 80 mol% SiO_2, this relationship appears to fail and the viscosity at T_g increases significantly (i.e. to 12 $\log_{10}$ Pa s). The failure of this viscosity-T_g relationship for SiO_2-rich melts needs to be investigated further.

The temperature dependent shear viscosity and the calorimetric T_g - heating-rate relationship determined via calorimetry measurements for matched cooling- and heating-rates for six natural silicate melts (Stevenson et al., 1995) are illustrated in Fig. 5.3. For the linear (Arrhenian) dependence of log_{10} viscosity on inverse temperature shown in Fig. 5.3, the slope of log_{10} heating-rate - inverse T_g data is the same as the log_{10} viscosity - inverse temperature data. Webb and Knoche (1996) have observed the same equivalence for 6 granitic melt composition in the case where a curved (non-Arrhenian) relationship is required to describe the dependence of log_{10} viscosity and log_{10} heating-rate upon inverse temperature. The constant required in Eqn. 5.2 to move the enthalpy relaxation data of Stevenson et al. (1995) and Webb and Knoche (1996) up to the shear viscosity curves ranges from 9.6 to 10.8. As G_∞ is not a constant independent of composition but varies from 5-35 GPa the value of the "constant" in Eqn. 5.2 is expected to vary by 0.9, slightly less than the 1.2 variation found experimentally. Scherer (1984) found the constant to be 11.3 for his melts as he defined T_g to occur at the onset of the peak in the heat capacity curve (this is a higher viscosity than that at the peak in the heat capacity curve).

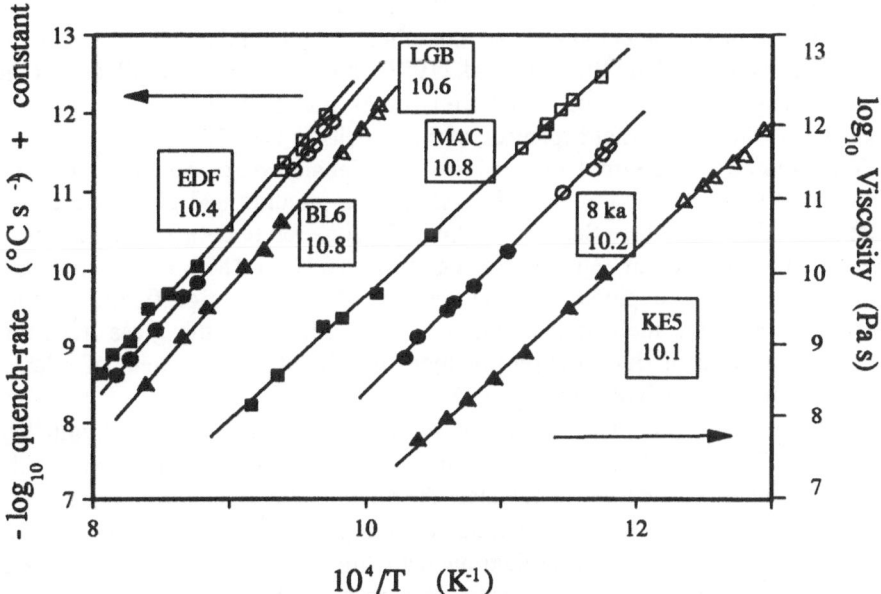

Fig. 5.3. Shear viscosity of six natural melts as a function of inverse temperature (solid symbols); together with the T_g values determined from enthalpy relaxation measurements for different quench-rates (hollow symbols). The constant required in Eqn. 5.2 is indicated for each melt. (redrawn after Stevenson et al. 1995)

5.2 Enthalpy and Volume

The density of silicate melts during rock-forming processes is a critical factor in the efficiency of crystal-melt fractionation, which ultimately gives rise to the diversity of igneous rocks. Accordingly, the experimental determination of melt densities and thermal expansivity has been the subject of considerable investigation in the geological sciences (Bottinga and Weill, 1970; Nelson and Carmichael, 1979; Mo et al., 1982; Stein et al., 1986; Lange and Carmichael, 1987; Dingwell et al., 1988). These investigations have been conducted at low viscosities in the superliquidus regime where problems of crystallisation do not occur. Igneous petrogenesis is, however, an overwhelmingly subliquidus subject. The processes leading to both erupted and intruded products of igneous fractionation occur at temperatures down to the solidus, and in disequilibrium cases, at subsolidus temperatures. Thermal expansivity data is therefore required at these low temperature/high viscosity conditions. Thermal expansivity data for silicate melts are required also as thermodynamic input for the calculation of physical properties such as derivation of melt compressibility from fusion curves of minerals, and the reduction of adiabatic velocity data to isothermal conditions.

Extrapolation of experimentally determined models of melt density to temperatures of geological interest are complicated by the large uncertainty in the thermal expansivity data. Bottinga et al. (1983) showed in their compilation of the thermal expansion data for Na_2O-SiO_2 melts that there is up to 20% error associated with the thermal expansion of these melts at 1700 K.

Two sources of thermal expansivity and density data are liquid density determination using the double-bob Archimedean method and glassy data from scanning dilatometry (see Sect. 4.3). The high temperature data are often limited by restricted ranges of temperature i.e. from 1673 to 1873 K. Dilatometric measurements of expansivity include the effects of significant viscous deformation of the melt as the sample is heated through the glass transition region. This viscous deformation precludes the direct measurement of relaxed liquid thermal expansivity (see Fig. 5.4).

Assuming the equivalence of relaxation parameters of volume and enthalpy, Webb et al. (1992) have used scanning calorimetry data to reconstruct the scanning dilatometry curves to calculate the volume and temperature dependence of volume ~50 K above the glass transition temperature. These low temperature data can then be used in combination with high temperature density data to interpolate volume and thermal expansivity across ~1000 K, thus filling in the gap in density data for silicate melts illustrated in Fig. 4.7.

The derivation of liquid expansivity and volume data from calorimetric and dilatometric measurements is based on the principles of structural relaxation in silicate melts. The general aspects of structural relaxation in silicate melts and their influence on viscosity, and density, have been discussed by Dingwell and Webb (1989; 1990) and Webb (1992a). The theory of the procedure for obtaining data on the molar expansivity of relaxed liquids from a combination of scanning calorimetry and dilatometry has been presented in full by Webb et al. (1992) and is reviewed below.

5.2.1 Theory

When a silicate glass is heated across the glass transition region a time-dependent response of its physical properties occurs. The unrelaxed, glassy values of volume and enthalpy are replaced by equilibrium, liquid values over a finite period of time. Quantitative models of structural relaxation in melts have been constructed (Narayanaswamy 1971, Moynihan et al., 1976; Scherer, 1984) to reproduce the details of the time-dependent response of melt properties in the glass transition interval. The models are completely general, describing the response of property P as a function of previous cooling-rate (from the relaxed state) and experimental heating-rate (from the unrelaxed state).

The physical properties of a silicate melt depend upon the ambient temperature T and the configuration or structure of the melt. Silicate glasses quenched from liquids preserve a configuration that can be approximated to the equilibrium structure of the liquid at some fictive temperature, T_f (Tool and Eichlin, 1931). To describe in general, the relaxed (liquid) or unrelaxed (glassy) properties of a silicate melt we need to specify the temperature and the fictive temperature of the melt. For a liquid, the structure is in equilibrium and thus T_f equals T (see Fig. 2.3). Upon cooling of the liquid into the glass transition region, the structure of the melt begins to deviate from equilibrium, i.e. T_f deviates from T. This deviation ultimately results in a temperature independence of T_f at low temperatures corresponding to the frozen structure of the glassy state. Upon subsequent reheating through the glass transition interval, the value of T_f once again assumes that of T and liquid values of melt properties are observed. The path of the value of the property taken during reheating is, however, different from that observed during cooling. Due to the finite rate of equilibration available for relaxation at the onset of the glass transition region there is an overshoot in the transient value of the melt property (i.e., the fictive temperature of the structure is lower than the melt temperature, $T_f < T$ (Dingwell and Webb, 1990)).

The temperature derivative of the physical properties of a glass and a liquid (e.g., molar heat capacity (dH/dT) and molar thermal expansivity (dV/dT)) can be used to describe the temperature-derivative of the fictive temperature. To do this, the temperature-derivative of any property in the glass transition interval (e.g., enthalpy, volume) is normalised with respect to the temperature-derivative of the liquid and glassy properties. This normalised temperature-derivative, which is equal to dT_f/dT, must equal zero for the glass (T_f is constant) and 1 for the liquid (T_f equals T).

The temperature-derivative of the fictive temperature T_f at a temperature T' is related to the temperature dependence of a macroscopic property P by;

$$
\left.\frac{dT_f}{dT}\right|_{T'} = \frac{\left[(\partial P/\partial T) - (\partial P/\partial T)_g\right]_{T'}}{\left[(\partial P/\partial T)_e - (\partial P/\partial T)_g\right]_{T_f}}
\tag{5.3}
$$

where the subscripts "e" and "g" are for the liquid (equilibrium) and the glassy values of the property (Moynihan et al. 1976).

In the present studies, enthalpy H, and volume V take the place of the general property P in Eqn. 5.3. Assuming the equivalence in relaxation time and activation energy for volume and enthalpy relaxation for melts with identical thermal histories (Webb, 1992a), Eqn. 5.3 can be rewritten as (Webb et al., 1992);

$$
\frac{c_p(T') - c_{pg}(T')}{c_{pe}(T_f) - c_{pg}(T_f)} = \left.\frac{dT_f}{dT}\right|_{T'} = \frac{\left(\dfrac{dV(T)}{dT} - \dfrac{dV_g(T)}{dT}\right)_{T'}}{\left(\dfrac{dV_e(T)}{dT} - \dfrac{dV_g(T)}{dT}\right)_{T_f}} .
\tag{5.4}
$$

Thus, in the glass transition region, the behaviour of any temperature dependent property of a melt can be predicted from the known behaviour of another temperature dependent property if the relaxation of the two properties is equivalent. The description of melt properties in the glass transition region thus no longer requires an algorithm to describe the temperature dependence of the fictive temperature. In the above equation, which relates c_p and thermal expansivity dV/dT, the only unknown parameter is the thermal expansivity of the relaxed liquid at temperature T' in the glass transition interval.

Due to the lack of relaxed thermal expansivity data (viscous flow of the melt above T_g results in the inability to determine volume as a function of temperature), the liquid molar thermal expansivity can be calculated from the dilatometric trace by normalising both the scanning calorimetric and dilatometric data;

$$
P'(T) = \frac{P(T) - P_g(T)}{P_p - P_g(T)}
\tag{5.5}
$$

where the subscripts "p" and "g" refer to peak and unrelaxed, glassy values. The relaxed value of thermal expansivity can now be generated from the peak and extrapolated glassy values of thermal expansivity; the volume and coefficient of volume thermal expansion α_v [1/V(dV/dT)] of the melt can be calculated as illustrated in Fig. 5.4.

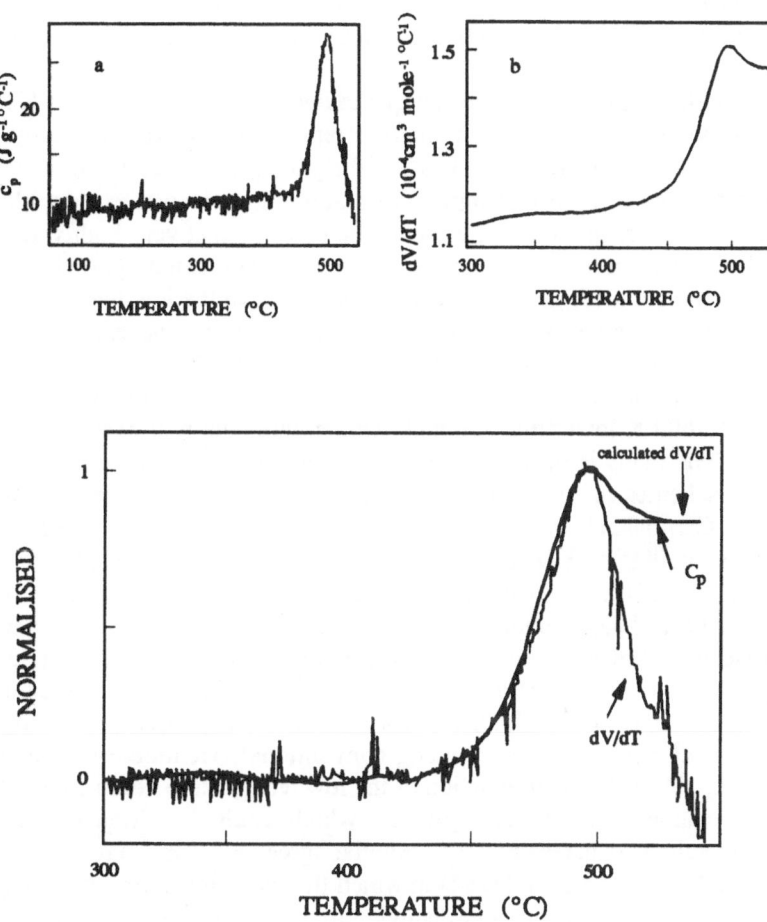

Fig. 5.4. a Scanning calorimetry and **b** dilatometry determinations of the heat capacity and thermal expansion of $Na_2Si_3O_7$ melt in the glass transition interval for melts which have been cooled at 5 K min^{-1} and reheated at 5 K min^{-1}. The onset of viscous deformation occurs above the peak temperature for this heating-rate. **c** Comparison of calorimetric and dilatometry traces normalised by Eqn. 5.4. (redrawn after Webb et al., 1992)

It must be stressed that the above method can only be applied to calorimetric and dilatometric data obtained on the same sample using identical thermal histories and heating-rates. It is only this internal consistency that permits the use of the assumption of the equivalence in behaviour of the enthalpy and volume relaxation. Small changes in composition or fictive temperature of the melt can strongly influence relaxation behaviour. In the case of "albite" ($NaAlSi_3O_8$) composition melt a 1 wt% variation in Na_2O was found to result in a 100 K variation in T_g obtained from both dilatometry and calorimetry (Knoche et al., 1992b).

Before the development of this technique to determine volume and dV/dT of a melt in the vicinity of the glass transition temperature there were only high temperature density data available. Most of the geological processes in which we are interested occur at low temperatures and high viscosities (double-bob Archimedean density measurements cannot be performed at very high viscosities as it is difficult to immerse the bob in the melts). The usual approach was to linearly extrapolate the high temperature density data to the low temperatures of interest. It has, however, now been shown that there is *not necessarily* a linear temperature dependence of volume in silicate melts (Knoche et al., 1992a,b). For the case of diopside composition ($CaMgSi_2O_6$) melt a linear extrapolation to a temperature ~600 K lower than the high temperature data will over-estimate the volume by 3%. (This is outside the ≤1% error associated with multicomponent calculation schemes for the density of silicate melts which are currently in use (Lange and Carmichael, 1987)). This would then have serious consequences for the further modelling of the sinking or floating of crystals in these melts.

The reduction of the behaviour of melt properties in the glass transition region to a universal set of parameters that describes the evolution of fictive temperature during reheating allows the extension of the dilatometric data into the liquid state and the calculation of volume and dV/dT for the supercooled melt. This method yields excellent agreement with the density and molar expansivity data obtained on $Na_2Si_3O_7$ at high temperature using conventional Archimedean techniques (Webb et al., 1992). The extrapolation of the low temperature volume and dV/dT was in agreement with the straight line which could be fitted to the high temperature volume data (see Fig. 5.5). Similar agreement was found in the Na_2O-SiO_2 system by Knoche et al. (1994) in which the low temperature data could be successful extrapolated to the high temperature data. For the case of the Na_2O-SiO_2 system, linear extrapolation of the high temperature data successfully predicted the low temperature volume. This simple relationship is however, not always the case for silicate melts (see Sect. 5.2.3 and 5.2.4). The combination of methods reduces the error in molar expansivity to 5% for melts with less than 90 mol% SiO_2. In melts with more than 90 mol% SiO_2 the error in α_v is 15%.

For all of the melt compositions discussed below, the T_g obtained from calorimetry and dilatometry data are identical within error of the measurements, and the calculated volume and dV/dT for the supercooled melt match with the extrapolated high temperature volumes in both the case of linear and non-linear temperature dependence of volume. This technique for determining the volume and dV/dT of a melt at a temperature just above the glass transition can be combined with density data at superliquidus temperature obtained from conventional Archimedean double-bob techniques to give density data from the supercooled melt up to super-liquidus temperatures. This results in knowledge of the density of melts which are involved in high viscosity melt fractionation and crystallisation processes.

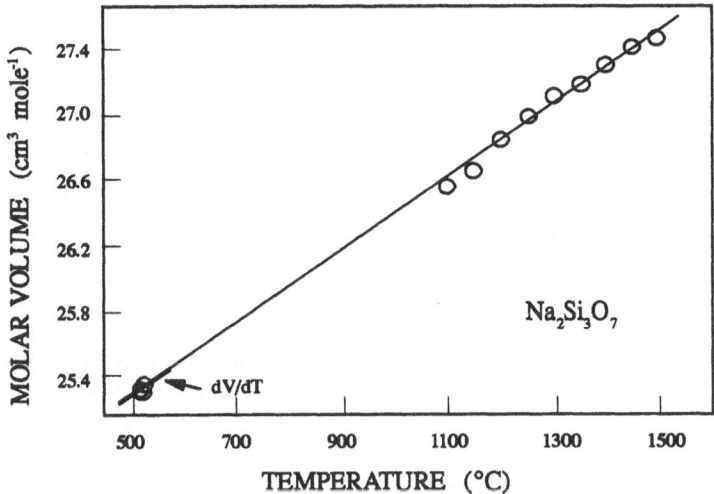

Fig. 5.5. Linear fit to the liquid volume data for $Na_2Si_3O_7$ obtained from high-temperature Archimedean and low-temperature dilatometry/calorimetry measurements. The thermal expansivity obtained from the combination of dilatometry and calorimetry constrain the slope of the fitted line at low temperatures. (redrawn after Webb et al., 1992)

5.2.2 Granitic Melts

The above technique to determine the volumes of a supercooled melts at a temperate just above the glass transition temperature has been used to determine the density of fluorine and boron-bearing melts of granitic composition. The addition of 7 mol% F_2O_{-1} (Dingwell et al., 1993a) and 8 mol% B_2O_3 results in a 3% decrease in melt density at 1023 K with respect to the density of the original granitic composition HPG8 (Knoche et al., 1992). Knoche et al. (1995) have used this technique to determine the change in molar volume of the same sub-aluminous granitic melt upon addition of alkali- and alkaline earth-oxides, and a number of high field-strength oxides.

The decrease in density due to the addition of fluorine and boron, along with the viscosity-reducing effect of boron (Dingwell et al., 1992) and fluorine on granitic melts (Dingwell et al., 1993a) should significantly accelerate processes of crystal-melt fractionation and facilitate the evolution of extremely fractionated igneous systems. The reduction in melt density will increase the density contrast between leucogranitic and pegmatitic melts and their equilibrium crystalline products. The decrease in melt density due to the addition of boron or fluorine will therefore accelerate crystal-melt fractionation processes that rely on density contrast. The effect of boron and fluorine on liquid density is complemented by the strong reduction in viscosity associated with their incorporation in the melt. This decrease in density and viscosity should combine to extend the igneous stage of boron-rich granitic and pegmatitic systems to lower temperatures and more extreme chemical compositions.

5.2.3 Thermal Expansion of GeO$_2$ and SiO$_2$ Melts

It is difficult to measure the coefficient of thermal expansion of SiO$_2$ melt. Bacon et al. (1960) determined the density of SiO$_2$ melt from 2223-2473 K and found α_v = 108×10^{-6} K^{-1}. This value, however, is in disagreement with the value of zero, determined from the calculation of the partial molar expansivity of silicate melts (Lange and Carmichael, 1990). The theoretical argument for a small (~zero) value of α_v for SiO$_2$ melt posed by Richet et al. (1982) was based on an incorrect value of the compressibility of molten SiO$_2$ (see Dingwell et al., 1993d). Thus the major theoretical objection to Bacon et al.'s data is removed. GeO$_2$ is commonly used as an analogue of SiO$_2$ in studies of the structure and properties of glasses and melts (Richet, 1990). For studies of the liquid state, GeO$_2$ provides the considerable advantage that the temperature required to achieve the relaxed liquid response of amorphous GeO$_2$ is ~600 K lower than that of SiO$_2$, at any given frequency. The lower temperature of structural relaxation in GeO$_2$ (at ~853 K versus 1450 K for SiO$_2$ at 10^{-2} Hz) brings studies of the glass transition of this single component network-structure liquid within the temperature range of operation of very precise scanning methods of dilatometry and calorimetry.

The volume and dV/dT of GeO$_2$ melt has been determined at 933 K by Dingwell et al. (1993d) from the combination of dilatometry and calorimetry data. The low-temperature liquid volume and expansivity of GeO$_2$ have been combined with high temperature densitometry determinations of the liquid volume of GeO$_2$ by Sekiya et al. (1980) to yield a temperature-volume relation for GeO$_2$ melt from 933 to 1673 K. GeO$_2$ shows a strongly temperature-dependent liquid molar expansivity (see Fig. 5.6), decreasing from 20×10^{-4} cm mol^{-1} K^{-1} to 2×10^{-4} cm mol^{-1} K^{-1} with increasing temperature. The coefficient of volume thermal expansion (α_v) decreases from 76×10^{-6} K^{-1} to 3×10^{-6} K^{-1} with increasing temperature. A qualitatively similar volume-temperature relationship has been observed previously in liquid B$_2$O$_3$ (see Fig. 5.6).

This data on the changes in heat capacity (Δc_p) and thermal expansivity ($\Delta \alpha_v$) across the glass transition, the molar volume (V) at the glass transition and the glass transition temperature (T$_g$) can be combined with literature-derived estimates of the compressibility of GeO$_2$ glass and liquid to estimate the values of the Prigogine-Defay ratio Π (Dingwell et al., 1993c; $\Pi = \Delta\beta \, \Delta c_p / [(\Delta\alpha_v)^2 \, T_g \, V]$

for the change in compressibility between the liquid and the glass, $\Delta\beta$), and the thermal Grüneisen parameter γ_{th}, for GeO_2 melt. The geophysical parameters Π and γ_{th} for the GeO_2 melt and silicate melts in general allow the estimation of the coefficient of volume thermal expansion of SiO_2 melt.

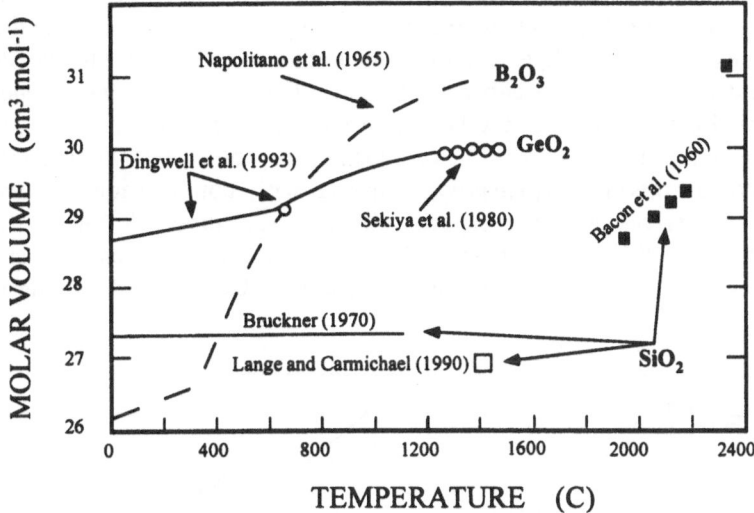

Fig. 5.6. The molar volume for SiO_2, GeO_2 and B_2O_3 glasses and liquids as a function of temperature. (redrawn after Dingwell et al., 1993)

5.2.4 Albite-Anorthite-Diopside Melts: $NaAlSi_3O_8$ - $CaAl_2Si_2O_8$ - $CaMgSi_2O_6$

This system has been historically applied as a model for basalt petrogenesis (e.g. phase equilibria, Bowen, 1915; Schairer and Yoder, 1960; Osborn and Tait, 1952; Kushiro, 1976: thermochemistry; Weill et al., 1980; Navrotsky et al., 1980; 1989: viscosity; Scarfe et al., 1983; Scarfe and Cronin, 1986, Tauber and Arndt, 1987: ultrasonic wave velocities; Rivers and Carmichael, 1987: shock wave equation of state; Rigden et al., 1988;1989).

The temperature-dependent thermal expansivities of melts in the system albite-anorthite-diopside have been determined on glassy and liquid samples using a combination of methods. Superliquidus volumes were determined using double Pt bob Archimedean densitometry at temperatures up to 2073 K. Supercooled liquid volumes and molar thermal expansivities were determined using scanning calorimetric and dilatometric measurements of properties in the glass transition region and their behaviour at the glass transition. The extraction of low temperature liquid molar expansivities from dilatometry/calorimetry is based on the equivalence of the relaxation of volume and enthalpy at the glass transition using the method developed by Webb et al. (1992).

The resulting data for liquid volumes near glass transition temperatures (1033-1193 K) and at superliquidus temperatures (1673-2073 K) are combined to yield thermal expansivities over the entire supercooled and stable liquid range. The volume thermal expansivities α_v are, in general, temperature-dependent (Knoche et al., 1992a,b). The temperature dependence of thermal expansivity in the melt increases from anorthite to albite to diopside composition (Fig. 5.7). The thermal expansivity of anorthite is essentially temperature-independent whereas that of diopside decreases by ~50% between 1073 and 1773 K; with the consequence that the thermal expansivities of the liquids in the diopside-albite-anorthite system converge at high temperatures. The composition-dependence of the temperature-derivative of thermal expansivity contrasts with composition-independent temperature-derivative of heat capacity obtained from drop calorimetry (Richet and Bottinga, 1986) and estimates of the temperature dependence of entropy derived from calorimetry and viscometry studies (Richet, 1984; Richet et al., 1986).

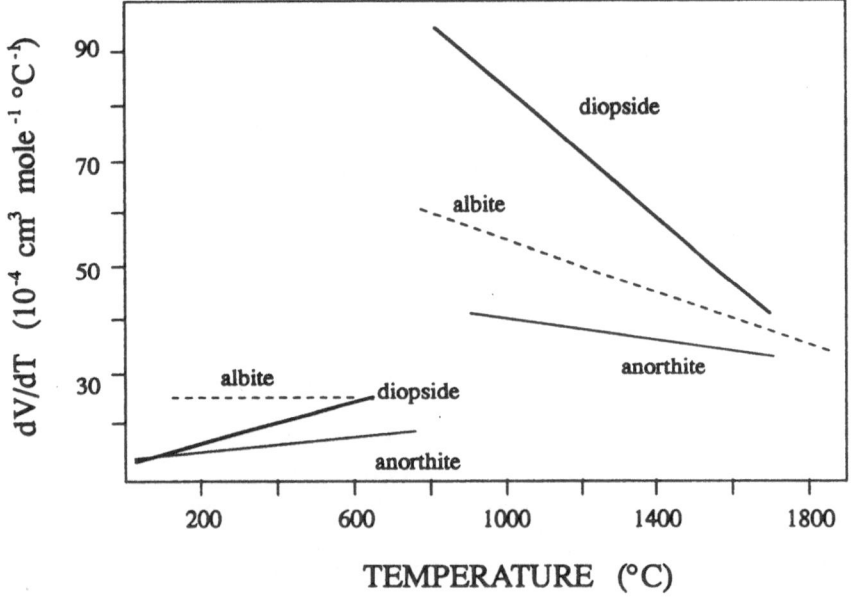

Fig. 5.7. dV/dT for albite, anorthite and diopside composition melts from room temperature to superliquidus temperatures. (redrawn after Knoche et al., 1992b)

Taniguchi (1989) has also investigated the thermal expansion of melts in the system diopside - anorthite using dilatometry. He used the peak values of thermal expansivities (the maximum in the dV/dT curve) for the liquid dV/dT at the glass transition; with the result that the values obtained were consistently too high to reproduce the high temperature volumes of his liquids. His data illustrates, for example, the contrast between a "peak" coefficient of thermal expansion of 240× 10^{-6} K^{-1} versus a coefficient of thermal expansion of 125×10^{-6} K^{-1} obtained from a polynomial fit to density extrapolated to the "peak" temperature, for diopside. In the study of Knoche et al. (1992a), the measured thermal expansivities near T_g are consistent with the values obtained from the fits to the volume. The measured and fitted values of the coefficient of thermal expansion for diopside at 1083 K are 125×10^{-6} K^{-1} and 123×10^{-6} K^{-1} respectively (Knoche et al., 1992a). It must be stressed that there is good agreement between the expansivity data obtained from dilatometry/calorimetry and those obtained from the fits to the combined (high- and low-temperature) melt volume data. This agreement removes the discrepancy posed by the Taniguchi (1989) data and improves confidence in the prediction of low temperature melt densities.

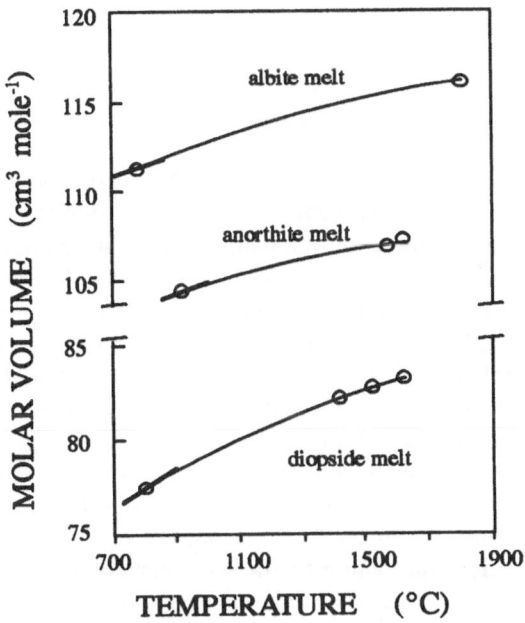

Fig. 5.8. Relaxed liquid volume data from dilatometry and densitometry for albite, anorthite and diopside composition melts. The thicker lines at low temperature indicate the calculated low temperature thermal expansivity (dV/dT). (redrawn after Knoche et al., 1992b)

An inspection of Fig. 5.8 indicates that extrapolation of diopside volume data either up or down temperature from the segments of the volume curve covered by either method, will result in serious overestimates of liquid volume. The extrapolation of high temperature volumes and expansivity to low temperature yields a 3% error. This discrepancy lies outside the ~1% accuracy of multicomponent calculation schemes presently available (e.g., Lange and Carmichael, 1987). Clearly, a much more complete investigation of the temperature-dependence of liquid expansivity is required to incorporate this deviation in such schemes. The improved precision of this method sets the stage for a detailed investigation of the composition dependence of expansivity in silicate melts. This type of measurement reduces the error in thermal expansivity measurements and allows measurement of volume at temperatures relevant to high viscosity processes and allows the interpolation through the experimentally inaccessible temperature interval below the liquidus that is so important for igneous petrogenesis. The gap in data as a function of temperature observed in Fig. 4.7 is essentially removed with the use of the present technique.

5.3 Equilibrium and Non-Equilibrium T_g

The glass transition temperatures discussed here are obtained by scanning calorimetry and dilatometry, and shear viscosity measurements. The relaxation of the physical properties volume and enthalpy and the measured glass transition temperatures are therefore being observed as a function of temperature in melts whose structure is not in equilibrium with the ambient temperature. With a consistent definition of T_g, the same thermal history and heating rate for each melt and accurate determinations of T_g Webb and Knoche (1996) have shown that contrary to the observations of Scherer (1986) and Knoche at al. (1992a) there is a one order of magnitude range in viscosity for 70 silicate melts at the T_g determined from such temperature scanning measurements. At high viscosities, one order of magnitude variation in viscosity is equivalent to a 40 K range in temperature. Therefore, based on the high quality of the relaxation data now available it is now clear that although the equivalences in relaxation discussed here are still *applicable for each specific melt composition, they are not strictly true between melt compositions.*

The T_gs determined by scanning through temperature are not equivalent to a constant viscosity independent of composition because the measurements are performed upon melts *whose structure is not in equilibrium with the ambient temperature.* For fixed cooling- and heating-rates, the viscosity at which the peak in the heat capacity curve occurs for melts of different composition, is affected by the differences in the energy of enthalpy relaxation in the glass transition region, and also the different relaxation time distributions as a function of melt composition. The effect of the variation in energy term is modelled in Fig. 5.9 for two melts with different energies required to describe the temperature dependence of their viscosities. The heat capacity curves A and B are modelled for a constant heating-rate, from the non-equilibrium to the equilibrium condition. If for each melt, the relaxation begins at the same viscosity of 10^w Pa s, the viscosity at the peak in the c_p curve is different for the two melts – 10^x Pa s, for melt #1 and 10^y Pa s for melt #2. Such a phenomenon is due to the larger energy for viscous flow (and enthalpy relaxation and volume relaxation) for melt #2 than for melt #1. This is an over-simplified model of the relaxation process but illustrates the cause of the expected differences in viscosity at the peak in the property curve as a function of melt composition. The introduction of a non-Arrhenian viscosity curve would further shift the peak position.

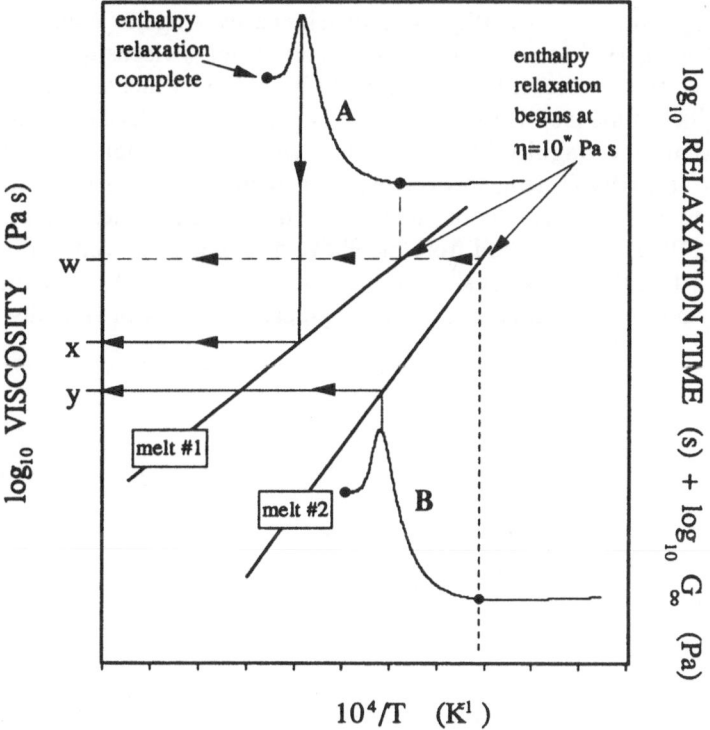

Fig. 5.9 The modelled viscosity (and relaxation time) occurring at the peak in the heat capacity curves as a function of the energy term required to describe the temperature dependence of viscosity.

The observed variations in viscosities at T_g and relaxation times in scanning determinations of the glass transition are due to the melt structure changing from the non-equilibrium state to being in equilibrium with the ambient temperature. In order to resolve the question whether the ± 0.5 $\log_{10}$ unit difference observed between some of the relaxation times obtained from NMR, shear flow, enthalpy relaxation, volume relaxation and light scattering measurements is real, it is necessary to determine relaxation times at a constant temperature as a function of the rate of perturbation and measurement of the melt response (see Fig. 2.2). This type of measurement will result in the observation of relaxation in a melt whose structure is in equilibrium with the ambient temperature - without the smearing out effect illustrated in Fig. 5.9.

Based upon the compilation of data presented by Webb and Knoche (1996) there does not appear to be any simple trend in the viscosity at T_g as a function of melt composition; nor as a function of the parameters NBO/T or fragility. The NBO/T ratio is a parameter commonly used to describe the structure of melts: that is the ratio of non-bridging oxygens per tetrahedrally bonded atom (see Mysen et al., 1984). The term "fragility" was developed by Angell (Wong and Angell, 1976; Angell, 1991) to describe the curvature of $\log_{10}$ viscosity as a function of reciprocal temperature. If $\log_{10} \eta$ can be fit by a straight line as a function of $1/T$, the fragility of the melt is low. If a curve is required to describe the $\log_{10} \eta$ - $1/T$ relationship, the fragility of the melt is high - and the melt is "strong". Both fragility and NBO/T tend to decrease with increasing SiO_2 content, but there appears to be no unique relationship between these two parameters (see Webb and Knoche, 1996). As expected from the above discussion, the constant required in Eqn. 5.2 to calculate viscosity from the rate dependent T_g determined from enthalpy relaxation measurements varies sympathetically with viscosity at T_g.

6 Summary

It has been shown that the glass transition temperatures observed for enthalpy-, volume- and shear-relaxation are related to the lifetime of the Si-O bonds in the melt. This structural relaxation is the slowest possible occurring in silicate melts at a specific temperature. There are no observations of slower relaxations due to the movement of polymers in the melt. All of the observations to date indicate that enthalpy-, volume- and shear-relaxation can be described by the same form of equation, with the same relaxation time and the same activation energy.

There are however, faster structural changes occurring in the melt. These are the movement of the other components through the melt; specifically the alkalies which move much faster than the Si or O ions. These relaxations will be observed at the same temperature as in the present studies, at frequencies orders of magnitude greater than those discussed here. These relaxations are observed in diffusion and electrical measurements. (The glass transition discussed here - that associated with the slowest structural relaxation - is the "glass transition with respect to Si and O"; while there are other glass transitions with respect to Na, and Al etc.; as it is possible to observe the rheology of a melt on a timescale faster than that of Na motion [unrelaxed with respect to Na movement], or a timescale longer than that of Na motion [relaxed with respect to Na movement].) For a full description of the structure of the melt it is necessary to investigate all of the possible structural relaxations occurring in the melt. The physical effects of these fast relaxations are, however, different to that of the Si-O relaxation, especially in the case of deformation. The shear attenuation due to alkali motion observed by Zdaniewski et al. (1979) using torsion techniques is three orders of magnitude less than that observed for the Si-O relaxation.

Understanding the fundamental structural cause of the glass transition, and the effects of the glass transition on melt rheology, there are a number of different directions of further investigation available:

First there is the effect of pressure. All of the studies mentioned here have been performed at one atmosphere pressure. While a large volume of melt exists on or near the Earth's surface, there are huge volumes of melt at pressures up to 1 GPa in the Earth's crust, and partially molten material up to pressures of 10 GPa in the asthenosphere (Ringwood, 1969). The basic questions of the effect of pressure on viscosity and density remain essentially unanswered. Direct measurements of these properties at high pressures and temperatures are very difficult and have large uncertainties associated with them. The measurement of wave velocities through melts, thermal expansion and differential thermal analysis measurements in melts as a function of temperature at high pressure will, however, result in the determination of compressibility and the estimation of viscosity from ultrasonic measurements; and the glass transition temperature and therefore viscosity as a function of heating-rate from the thermal expansion and heat capacity measurements; as illustrated in Sect. 3, 4 and 5.

Second there is the question of the rheology of natural lavas and magmas. The methods discussed here for direct measurement of melt rheology are also applicable to the study of crystal- and bubble-bearing melts, whether they are synthetic compositions or natural glasses (Bagdassarov et al., 1994; Dingwell, 1995; Lejeune and Richet, 1995; Stevenson et al., 1995; 1996). As a large amount of work has now been done on the rheology of crystal- and bubble-free melts it is possible to separate the effects of melt composition and temperature from the presence or absence of crystals and bubbles.

Third, there is the possibility to determine the thermal history of natural glasses from the glass transition temperature determined as a function of heating-rate in either thermal expansion measurements or heat capacity measurements. Using the algorithm of Narayanaswamy (1971) or the Adam-Gibbs algorithm of Crichton and Moynihan (1988) to describe the temperature dependence of the fictive temperature T_f, it is possible to model the position, height and width of the c_p peak as a function of cooling- and heating-rate of the melt. In the case of natural glasses which are the products of geological processes (i.e. lava flows) we can calculate the cooling-rate of the melt. This can also be used as a tool to investigate the complexity of the cooling history of a glassy deposit (see Martens et al., 1987; Wilding et al., 1995; Wilding et al., 1996).

The structure of melts need to be addressed further. The lifetime of Si-O bonds can be determined using a variety of techniques. There has been no detailed study of the effects of composition (i.e. the addition of aluminium, or variation in the network-forming and network modifying atoms) on the melt structure either by observing the effects on the lifetime of Si-O bonds or by observing the lifetime of the other bonds in the melt. The bulk modulus and density measurements can be used to indicate co-ordination changes occurring in the melt as a function of composition, or simply changes in the strength of bonds as the composition is changed. Density and heat capacity measurements as a function of composition indicate whether there is ideal or non-ideal mixing. Determination of the shear viscosity of melts over a large temperature range show how the structural relaxation time of the melt varies as a function of temperature and composition; which leads to the discussion of the structure of the melt. The observations made on the physical properties of silicate melts can be combined with spectroscopy measurements of the nearest neighbour atoms in order to obtain a more complete understanding of the melt structure.

References

Agee, C.B. and D. Walker, Static compression and olivine floatation in ultrabasic silicate liquids, *J. Geophys. Res.*, **93**, 3437-3449, 1988.

Anderson, O.L. and J.E. Nafe, The bulk modulus-volume relationship for oxide compounds and related geophysical problems, *J. Geophys. Res.*, **70**, 3951-3963, 1965.

Angell, C.A., Relaxation in liquids, polymers and plastic crystals - strong/fragile patterns and problems, *J. Non-Cryst. Sol.*, **131**, 13-31, 1991.

Askarpour, V., M.H. Manghnani and P. Richet, Elastic properties of diopside, anorthite, and grossular glasses and liquids: a Brillouin scattering study up to 1400K, *J. Geophys. Res.*, **98**, 17683-17689, 1993.

Babcock, C.L., Viscosity and electrical conductivity of molten glasses, *J. Am. Ceram. Soc.*, 329-342, 1934.

Bacon, J.F., A.A. Hasapis and J.W. Wholley, Viscosity and density of molten silica and high silica content glasses, *Phys. Chem. Glass.*, **1**, 90-98, 1960.

Bagdassarov, N.S. and D.B. Dingwell, A rheological investigation of vesicular rhyolite, *J. Volcanol. Geotherm. Res.*, **50**, 307-322, 1992.

Bagdassarov, N.S., D.B. Dingwell and S.L. Webb, Effect of boron, phosphorus and fluorine on shear stress relaxation in haplogranite melts, *Europ. J. Mineral.*, **5**, 409-425, 1993.

Bagdassarov, N.S., D.B. Dingwell and S.L. Webb, Viscoelasticity of crystal- and bubble-bearing rhyolite melts, *Phys. Earth Planet. Int.*, **83**, 83-99, 1994.

Bansal, N.P. and R.H. Doremus, Handbook of Glass Properties, Academic Press, London, 680p, 1986.

Birge, N.O. and S.R. Nagel, Specific-heat spectroscopy of the glass transition, *Phys. Rev. Lett.*, **54**, 2674-2677, 1985.

Bockris, J.O'M. and E. Kojonen, The compressibilities of certain molten alkali silicates and borates, *J. Am. Ceram. Soc.*, **82**, 4493-4497, 1960.

Bockris, J.O'M., J.D. MacKenzie and J.A. Kitchener, Viscous flow in silica and binary liquid silicates, *Trans. Farad. Soc.*, **51**, 1734-1748, 1955.

Bornhöft, H. and R. Brückner, Ultrasonic measurements and complex elastic moduli of silicate glass melts in the viscoelastic and viscous range, *Glastech. Ber.*, **67**, 241-254, 1994.

Bottinga, Y., Non-Newtonian rheology of homogeneous silicate melts, *Phys. Chem. Mineral.*, **20**, 454-459, 1994.

Bottinga, Y. and D.F. Weill, Density of liquid silicate systems calculated from partial molar volumes of oxide components, *Am. J. Sci.*, **269**, 169-182, 1970.

Bottinga, Y., D. Weill and P. Richet, Calculation of the density and thermal expansion coefficient of silicate liquids, *Bull. Mineral.*, **104**, 129-138, 1983.

Bowen, N.L., The crystallization of haplobasaltic, haplodioritic and related magmas, *Am. J. Sci.*, **40**, 161-185, 1915.

Brawer, S.; Relaxation in Viscous Liquids and Glasses, *Am. Ceram. Soc.*, 220p, 1985.

Bridgman, P.W., The compressibility of thirty metals as a function of pressure and temperature, *Proc. Amer. Acad. Arts Sci.*, **58**, 165-242, 1923.

Brückner, R., Structural aspects of highly deformed melts, *J. Non-Cryst. Solids*, **95-96**, 961-968, 1987.

Crichton, S.N. and C.T. Moynihan, Structural relaxation of lead silicate glass, *J. Non-Cryst. Solid.*, **102**, 222-227, 1988.

Day, D.E. and W.E. Steinkamp, In: Porai-Koshits, E.A. (ed.), *Stekloobraznoye Sostoyaniye*, 294-298, Leningrad, 1971.

DeBolt, M.A., A.J. Easteal, P.B. Macedo and C.T. Moynihan, Analysis of structural relaxation in glass using rate heating data, *J. Am. Ceram. Soc.*, **59**, 16-21, 1976.

Dingwell, D.B., Effects of structural relaxation on cationic tracer diffusion in silicate melts. *Chem. Geol.*, **82**, 209-216, 1990.

Dingwell, D.B., Density of some titanium-bearing silicate liquids and the compositional dependence of the partial molar volume of TiO_2, *Geochim Cosmochim Acta*, **56**, 3403-3407, 1992.

Dingwell, D.B., Relaxation in silicate melts: some applications, In: Structure, Dynamics and Properties of Silicate Melts, J.F. Stebbins, P.F. McMillan and D.B. Dingwell (eds), *Rev. Mineral.*, **32**, 21-66, 1995.

Dingwell, D.B. and S.L. Webb, Structural relaxation in silicate melts and non-Newtonian melt rheology in geologic processes, *Phys. Chem. Mineral.*, **16**, 508-516, 1989.

Dingwell, D.B. and S.L. Webb, Structural relaxation in silicate melts, *Europ. J. Mineral.*, **2**, 427-449, 1990.

Dingwell, D.B. and S.L. Webb, The fluxing effect of fluorine at magmatic temperatures (600-800°C): a scanning calorimetric study, *Am. Mineral.*, **77**, 30-33, 1992.

Dingwell, D.B., M. Brearley and J.E. Dickinson, Melt densities in the Na_2O-FeO-Fe_2O_3-SiO_2 system and the partial molar volume of tetrahedrally coordinated ferric iron in silicate melts, *Geochim. Cosmochim. Acta*, **52**, 2467-2475, 1988.

Dingwell, D.B., R. Knoche, S.L. Webb and M. Pichavant, The effect of B_2O_3 on the viscosity of haplogranitic liquids, *Am. Mineral.*, **77**, 457-461, 1992.

Dingwell, D.B., N. Bagdassarov, G. Bussod and S.L. Webb, Magma rheology, In: Mineralogical Association of Canada Short Course on Experiments at High Pressure and Applications to the Earth's Mantle, R.W. Luth (ed), 131-196, 1993a.

Dingwell, D.B., R. Knoche and S.L. Webb, The effect of fluorine on the density of haplogranite melt, *Am. Mineral.*, **78**, 325-330, 1993b.

Dingwell, D.B., R. Knoche and S.L. Webb, The effect of P_2O_5 on the viscosity of haplogranite liquid, *Europ. J. Mineral.*, **5**, 133-140, 1993c.

Dingwell, D.B., R. Knoche and S.L. Webb, A volume temperature relationship for liquid GeO_2 and some geophysically relevant derived parameters for network liquids, *Phys. Chem. Mineral.*, **19**, 445-453, 1993d.

Farnan, I and J.F. Stebbins, A high temperature ^{29}Si investigation of solid and molten silicates, *J. Am. Ceram. Soc.*, **112**, 32-39, 1990.

Farnan, I. and J.F. Stebbins, The nature of the glass transition in a silica-rich oxide melt, *Science*, **265**, 1206-1209, 1994.

Fontana, E.H. and W.A. Plummer, A viscosity-temperature relation for glass, *Am. Ceram. Soc.*, **62**, 367-369, 1979.

Gruber, G.J. and T.A. Litovitz, Shear and structural relaxation in zinc chloride, *J. Chem. Phys.*, **40**, 13-26, 1964.

Herzberg, C.T., Magma density at high pressure Part 1: The effect of composition on the elastic properties of silicate liquids, In: Magmatic Processes: Physicochemical Principles, B.O. Mysen (ed), *The Geochemical Society Special Publ.*, **1**, 25-46, 1987.

Herzfeld, K.F. and T.A. Litovitz, Absorption and Dispersion of Ultrasonic Waves, Academic Press, 535p, 1959.

Hess, K-U., D.B. Dingwell and S.L. Webb, The influence of excess alkalies on the viscosity of a haplogranitic melt, *Am. Mineral.*, **80**, 297-304, 1995a.

Hess, K-U., D.B. Dingwell and S.L. Webb, The influence of alkaline earth oxides (BeO, MgO, CaO, SrO, BaO) on the viscosity of a haplogranitic melt: systematics of non-Arrhenian behavior, *Europ. J. Mineral.*, **80**, 297-304, 1995b.

Illig, H., ABC Glas, Deutscher Verlag für Grundstoffindustrie, Leipzig, 1991.

Johnson, J.R., R.H. Bristow and H.H. Blau, Diffusion of ions in some simple glasses, *J. Am. Ceram. Soc.*, **34**, 165-172, 1951.

Knoche R., D.B. Dingwell and S.L. Webb, Temperature dependent thermal expansivities of silicate melts: the system anorthite-diopside, *Geochim. Cosmochim. Acta*, **56**, 689-699, 1992a.

Knoche R., D.B. Dingwell and S.L. Webb, Non-linear temperature dependence of liquid volumes in the system albite-anorthite-diopside, *Contrib. Mineral. Petrol.*, **111**, 61-73, 1992b.

Knoche R., S.L. Webb and D.B. Dingwell, A partial molar volume for B_2O_3 in haplogranitic melt, *Canadian Mineral.*, **30**, 561-569, 1992c.

Knoche, R., D.B. Dingwell, F.A. Seifert and S.L. Webb, Non-linear properties of supercooled liquids in the system Na_2O-SiO_2, *Chem. Geol.*, **116**, 1-16, 1994.

Knoche, R., D.B. Dingwell and S.L. Webb, Leucogranitic and pegmatitic melt densities: partial molar volumes for SiO_2, Al_2O_3, Na_2O, K_2O, Rb_2O, Cs_2O, Li_2O, BaO, SrO, CaO, MgO, TiO_2, B_2O_3, P_2O_5, F_2O-1, Ta_2O_5, Nb_2O_5 and WO_3, *Geochim. Cosmochim. Acta*, **59**, 4645-4652, 1995.

Kress V.C. and I.S.E. Carmichael, The compressibility of silicate liquids containing Fe_2O_3 and the effect of composition, temperature, oxygen fugacity and pressure on their redox states, *Contrib. Mineral. Petrol.*, **108**, 82-92, 1991.

Kress, V.C., Q. Williams and I.S.E. Carmichael, Ultrasonic investigations of melts in the system Na_2O-Al_2O_3-SiO_2, *Geochim. Cosmochim. Acta*, **52**, 283-293, 1988.

Kurkjian C.R., Relaxation of torsional stress in the transformation range of a soda-lime-silica glass, *Phys. Chem. Glass*, **4**, 128-136, 1963.

Kushiro I., Changes in viscosity and structure of melt of $NaAlSi_2O_6$ composition at high pressures, *J. Geophys. Res.*, **81**, 6347-6350, 1976.

Laberge N.L., V.V. Vasilescu, C.J. Montrose and P.B. Macedo, Equilibrium compressibilities and density fluctuations in K_2O-SiO_2 glasses, *J. Am. Ceram. Soc.*, **56**, 506-509, 1973.

Lange, R.L. and I.S.E. Carmichael, Densities of Na_2O-K_2O-CaO-MgO-FeO-Fe_2O_3-TiO_2-SiO_2 liquids: new measurements and derived partial molar properties, *Geochim. Cosmochim. Acta*, **51**, 2931-2946 1987.

Lange, R.A. and I.S.E. Carmichael, Thermodynamic properties of silicate liquids with emphasis on density, thermal expansion and compressibility, In: Modern Methods of Igneous Petrology, J. Nicholls and J.K. Russel (eds.), *Mineral. Soc. Amer.*, Washington, 25-64, 1990.

Lange, R.A. and A. Navrotsky, Heat capacities of TiO_2-bearing silicate liquids: Evidence for anomalous changes in configurational entropy with temperature, *Geochim. Cosmochim. Acta*, **57**, 3001-3011, 1993.

Lejeune, A-M. and P. Richet, Rheology of crystal-bearing silicate melts: an experimental study at high viscosities, *J. Geophys. Res.*, **100**, 4215-4229, 1995.

Li, J.H. and D.R. Uhlmann, The flow of glass at high stress levels, I. Non-Newtonian behaviour of homogeneous $0.08Rb_2O \cdot 0.92SiO_2$ glasses, *J. Non-Cryst. Solids*, **3**, 127-147, 1970.

Lillie, H.R., Viscosity-time-temperature relations in glass at annealing temperature, *J. Am. Ceram. Soc.*, **16**, 619-631, 1933.

Liu, S-B., J.F. Stebbins, E Schneider and A. Pines, Diffusive motion in alkali-silicate melts: an NMR study at high temperature, *Geochim. Cosmochim. Acta*, **52**, 527-538, 1988.

Macedo, P.B., J.H. Simmons and W. Haller, Spectrum of relaxation times and fluctuation theory: ultrasonic studies on an alkali-borosilicate melt, *Phys. Chem. Glass.*, **9**, 156-164, 1968.

Manghnani, M.H., C.S. Rai, K.W. Katahara and G.R. Olhoeft, Ultrasonic velocity and attenuation in basalt melt. In: Anelasticity in the Earth, F.D. Stacey, M.S. Paterson and A. Nicholas (eds), Geodynamics Series Volume 4, American Geophysical Union, 118-122, 1981.

Martens, R.M., M. Rosenhauer, H. Büttner and K. von Gehlen, Heat capacity and kinetic parameters in the glass transformation interval of diopside, anorthite and albite glass, *Chem. Geol.*, **62**, 49-70, 1987.

Maxwell, J.C., On the dynamical theory of gases, *Phil. Trans. Roy. Soc.*, **157**, 49-88, 1867.

Mazurin, O.V., Glass relaxation, *J. Non-Cryst. Solids*, **87**, 392-407, 1986.

Mazurin, O.V., M.V. Streltsina and T.P. Shvaiko-Shvaikovskaya, Handbook of Glass Data: Part C, Elsevier, Amsterdam, 1110p, 1987.

Miller, G.H., E.M. Stolper and T.J. Ahrens, The equation of state of molten komatiite, 1, shock wave compression to 36 GPa, *J. Geophys. Res.*, **96**, 11831-11848, 1991.

Mills, J.J., Low frequency storage and loss moduli of soda silica glasses in the transformation range, *J. Non-Cryst. Solids*, **14**, 255-268, 1974.

Mo, X., I.S.E. Carmichael, M Rivers and J. Stebbins, The partial molar volume of Fe_2O_3 in multicomponent silicate liquids and the pressure dependence of oxygen fugacity in magmas, *Mineral. Mag.*, **45**, 237-245, 1982.

Moynihan, C.T., AJ. Easteal, J. Wilder and J. Tucker, Dependence of the glass transition temperature on heating and cooling rate, *J. Phys. Chem.*, **78**, 2673-2677, 1974.

Moynihan, C.T., A.J. Easteal, M.A. DeBolt and J. Tucker, Dependence of fictive temperature of glass on cooling rate, *J. Am. Ceram. Soc.*, **59**, 12-16, 1976a.

Moynihan, C.T., A.J. Easteal, D.C. Tran, J.A. Wilder and E.P. Donovan, Heat capacity and structural relaxation of mixed-alkali glasses, *J. Am. Ceram. Soc.*, **59**, 137-140, 1976b.

Mysen, B.O., D. Virgo and F.A. Seifert, Redox equilibria of iron in alkaline earth silicate melts: relationships between melt structure, oxygen fugacity, temperature and properties of iron-bearing silicate melts, *Am. Mineral.*, **54**, 834-847, 1984.

Narayanaswamy, O.S., A model of structural relaxation in glass, *J. Am. Ceram. Soc.*, **54**, 491-498, 1971.

Narayanaswamy, O.S., Thermorheological simplicity in the glass transition, *J. Am. Ceram. Soc.*, **71**, 900-904, 1988.

Navrotsky, A., R. Hon, D.F. Weill and D.J. Henry, Thermochemistry of glasses and liquids in the system $CaMgSi_2O_6$-$CaAl_2Si_2O_8$-$NaAlSi_3O_8$, SiO_2-$CaAl_2Si_2O_8$-$NaAlSi_3O_8$ and SiO_2-Al_2O_3-CaO-Na_2O, *Geochim. Cosmochim. Acta*, **44**, 1409-1423, 1980.

Navrotsky, A., D. Ziegler, R. Oestrike and P. Maniar, Calorimetry of silicate melts at 1773K: measurements of enthalpies of fusion and of mixing in the systems diopside-anorthite-albite and anorthite-forsterite, *Contrib. Mineral. Petrol.*, **101**, 122-130, 1989.

Nelson, S.A. and I.S.E. Carmichael, Partial molar volume of oxide components in silicate liquids, *Contrib. Mineral. Petrol.*, **71**, 117-124, 1979.

Nowick, A.S. and B.S. Berry, Anelastic Relaxation in Crystalline Solids, Academic Press, New York, 1972.

Osborn, E.F. and D.B. Tait, The system diopside-forsterite-anorthite, *Am. J. Sci.*, Bowen Volume, 413-433, 1952.

Perez, J., A. Duperray and D. Lefevre, Viscoelastic behaviour of an oxide glass near the glass transition temperature, *J. Non-Cryst. Solid.*, **44**, 113-136, 1981.

Paris, E., D.B. Dingwell, F.A. Seifert, A. Mottana and C. Romano, An X-ray absorption study of Ti in Ti-rich silicate melts correlations with melt properties, *Terra Abstracts*, **5**, 521, 1993.

Poole, J.P., Viscosité à basse température des verres alcalino-silicatés, *Verres Refract.*, **2**, 222-228, 1948.

Provencano, V., L.P. Boesch, V. Volterra, C.T. Moynihan and P.B. Macedo, Electrical relaxation in $Na_2O.3SiO_2$ glass, *J. Am. Ceram. Soc.*, **55**, 492-496, 1972.

Rekhson, S.M, Measuring physical properties of glass in the glass-transition region, *Ceram. Bull.*, **68**, 1956-1962, 1989.

Rekhson, S.M., A.V. Bulaeva and O.V. Mazurin, Changes in the linear dimensions and viscosity of window glass during stabilization, *Inorg. Mater. (Engl. Trans.)*, **7**, 622-623, 1971.

Richet, P., Viscosity and configurational entropy of silicate melts, *Geochim. Cosmochim. Acta*, **48**, 471-483, 1984.

Richet, P., GeO_2 vs SiO_2: glass transitions and thermodynamic properties of polymorphs, *Phys. Chem. Mineral.*, **17**, 79-88, 1990.

Richet, P. and Y. Bottinga, Anorthite, andesine, wollastonite, diopside, cordierite and pyrope: thermodynamics of melting, glass transitions, and properties of amorphous phases, *Earth Planet. Sci. Lett.*, **67**, 415-432, 1984.

Richet, P. and Y. Bottinga, Heat capacity of aluminium-free liquid silicates, *Geochim. Cosmochim. Acta*, **49**, 471-486, 1985.

Richet, P. and Y. Bottinga, Thermophysical properties of silicate glasses and liquids, *Rev. Geophys.*, **24**, 1-25, 1986.

Richet, P. and Y. Bottinga, Rheology and configurational entropy of silicate melts, In: Structure, Dynamics and Properties of Silicate Melts, J.F. Stebbins, P.F. McMillan and D.B. Dingwell (eds), *Rev. Mineral.*, **32**, 67-93, 1995.

Richet, P. and D.R. Neuville, Thermodynamics of silicate melts: configurational properties, In: Thermodynamic Data, S. Saxena (ed), Springer, 132-160, 1992.

Richet P., Y. Bottinga, L. Denielou, J.P. Petitet and C. Tequi, Thermodynamic properties of quartz, cristobalite and amorphous SiO_2: drop calorimetry measurements between 1000 and 1800 K and a review from 0 to 2000 K, *Geochim. Cosmochim. Acta*, **46**, 2639-2658, 1982.

Richet, P., R.A. Robie and B.S. Hemingway, Low-temperature heat capacity of diopside glass (CaMgSi$_2$O$_6$): A calorimetric test of the configurational-entropy theory applied to the viscosity of liquid silicates, *Geochim. Cosmochim. Acta*, **50**, 1521-1533, 1986.

Rigden, S., T.J. Ahrens and E.M. Stolper, Shock compression of molten silicate: Results for a model basaltic composition, *J. Geophys. Res.*, **93**, 367-382, 1988.

Rigden, S., T.J. Ahrens and E.M. Stolper, High-pressure equation of state of molten anorthite and diopside, *J. Geophys. Res.*, **94**, 9508-9522, 1989.

Ringwood, A.E., Composition and evolution of the upper mantle, AGU Monograph, **13**, 1-17, 1969.

Rivers, M.L. and I.S.E. Carmichael, Ultrasonic studies of silicate melts, *J. Geophys. Res.*, **92**, 9247-9270, 1987.

Rosen, S.L., Fundamental Principles of Polymeric Materials, Wiley, New York, 346p, 1982.

Ryan, M.P. and J.Y.K. Blevins, The viscosity of synthetic and natural silicate melts and glasses at high temperatures at 1 bar (10^5 Pa) pressure and at higher pressures, 563p, 1987.

Sasabe, H., M.A. DeBolt, P.B. Macedo and C.T. Moynihan, Structural relaxation in an alkali-lime-silicate glass. In Proceedings of the 11th International Congress on Glass, Prague, Vol. 1, 339-348, 1977.

Sato, H. and M.H. Manghnani, Ultrasonic measurements of V_p and Q: relaxation spectrum of complex modulus of basalt melts, *Phys. Earth Planet. Int.*, **41**, 18-33, 1985.

Scarfe, C.M. and D.J. Cronin, Viscosity-temperature relationships at 1 atm of melts in the system diopside-albite, *Am. Mineral.*, **74**, 767-771, 1986.

Scarfe, C.M., D.J. Cronin, J.T. Wenzel and D.A. Kaufmann, Viscosity-temperature relationships at 1 atm in the system diopside-anorthite, *Am. Mineral.*, **68**, 1083-1088, 1983.

Scarfe C.M., B.O. Mysen and D. Virgo, Pressure dependence of the viscosity of silicate melts, In: Magmatic processes: physicochemical principles, *The Geochem. Soc. Special Publ.*, **1**, 59-67, 1987.

Schairer, J.F. and H.S. Yoder, The nature of residual liquids from crystallization, and data on the system nepheline-diopside-silica, *Am. J. Sci.*, **258A**, 273-283, 1960.

Scherer, G.W., Use of the Adam-Gibbs equation in the analysis of structural relaxation, *J. Am. Ceram. Soc.*, **67**, 504-511, 1984.

Scherer, G.W., Relaxation in Glass and Composites, Wiley, New York, 331p, 1986.

Secco, R.A., M.H. Manghnani and T-C. Liu, The bulk modulus-attenuation-viscosity systematics of diopside-anorthite melts, *Geophys. Res. Lett.*, **18**, 93-96, 1991.

Seddon, E., E.J. Tippert and W.E.S. Turner, *J. Soc. Glass. Techn.*, **16**, 450, 1932.

Sekiya, K., K. Morinaga and T. Yanagase, Physical properties of Na$_2$O-GeO$_2$ melts, *J. Jpn. Ceram. Soc.*, **88**, 367-373, 1980.

Siewert, R, Viskoelastische Relaxationsmessungen in Schmelzen der Zusammensetzung NaAlSi$_3$O$_8$, NaAlSi$_2$O$_6$, CaAl$_2$Si$_2$O$_8$ und CaMgSi$_2$O$_6$ mit Hilfe der Photonenkorrelations-Spektroskopie, PhD. Thesis, Georg-August-Universität, Göttingen, 130p, 1993.

Siewert, R. and M. Rosenhauer, Light scattering in jadeite melt: strain relaxation measurements by photon correlation spectroscopy, *Phys. Chem. Mineral.*, **21**, 18-23, 1994.

Simmons, J.H. and P.B. Macedo, Viscous relaxation above the liquid-liquid phase transition in some oxide mixtures, *J. Chem. Phys.*, **53**, 2914-2922, 1970.

Simmons, J.H., R.K. Mohr and C.J. Montrose, Non-Newtonian viscous flow in glass, *J. Appl. Phys.*, **53**, 4075-4080, 1982.

Spera, F.J., A. Borgia and J. Strimple, Rheology of melts and magmatic suspensions. 1. design and calibration of concentric cylinder viscometer with application to rhyolitic magma, *J. Geophys. Res.*, **93**, 10273-10294, 1988.

Stebbins, J.F., Dynamics and structure of silicate and oxide melts: nuclear magnetic resonance studies, In: Structure, Dynamics and Properties of Silicate Melts, J.F. Stebbins, P.F. McMillan and D.B. Dingwell (eds), *Rev. Mineral.*, **32**, 191-246, 1995.

Stebbins, J.F., S. Sen and I. Farnan, Silicate species exchange, viscosity, and crystallization in a low-silica melt: In situ high-temperature MAS NMR spectroscopy. *Am. Mineral.*, **80**, 861-864, 1995.

Stein, D.J. and F.J. Spera, Rheology and microstructure of magmatic emulsions: theory and experiments, *J. Volc. Geotherm. Res.*, **49**, 157-174, 1992.

Stein, D.J., J.F. Stebbins and I.S.E. Carmichael, Density of molten sodium aluminosilicates, *J. Am. Ceram. Soc.*, **69**, 396-399, 1986.

Stevenson, R.J., D.B. Dingwell, S.L. Webb and N.S. Bagdassarov, The equivalence of enthalpy and shear stress relaxation in rhyolitic obsidians and quantification of the liquid-glass transition in volcanic processes, *J. Volcanol. Geotherm. Res.*, **68**, 297-306, 1995.

Stevenson, R.J., D.B. Dingwell, S.L. Webb and T.G. Sharp, Viscosity of microlite-bearing rhyolitic obsidians: and experimental study, *Bull. Volcanol.*, **58**, 298-309, 1996.

Stokes, G.G., *Trans. Camb. phil. Soc.*, **8**, 287, 1845.

Taniguchi, H., Densities of melts in the system $CaMgSi_2O_6$-$CaAl_2Si_2O_8$ at low and high pressures, and their structural significance, *Contrib. Mineral. Petrol.*, **103**, 325-334, 1989.

Tauber, P. and J. Arndt, Viscosity and temperature relationship of liquid diopside, *Phys. Earth Planet. Int.*, **43**, 97-103, 1986.

Tauber, P. and J. Arndt, The relationship between viscosity and temperature in the system anorthite-diopside, *Chem. Geol.*, **62**, 71-81, 1987.

Tauke, J., T.A. Litovitz and P.B. Macedo, Viscous relaxation and non-Arrhenius behaviour in B_2O_3, *J. Am. Ceram. Soc.*, **51**, 158-163, 1968.

Tool, A.Q. and C.G. Eichlin, Variations caused in the heating curves of glass by heat treatment, *J. Am. Ceram. Soc.*, **14**, 276-308, 1931.

Watt, J.P., G.F. Davies and R.J. O'Connell, The elastic properties of composite materials, *Geophys. J. R. astr. Soc.*, **54**, 601-630, 1976.

Webb, S.L., Shear and volume relaxation in $Na_2Si_2O_5$, *Am. Mineral.*, **76**, 1449-1454. 1991.

Webb, S.L., Shear, volume, enthalpy and structural relaxation in silicate melts, *Chem. Geol.*, **96**, 449-458, 1992a.

Webb, S.L., Low-frequency shear and structural relaxation in rhyolite melt, *Phys. Chem. Mineral.*, **19**, 240-245, 1992b.

Webb, S.L. and P. Courtial, Compressibility of melts in the CaO-Al_2O_3-SiO_2 system, *Geochim. Cosmochim. Acta*, **60**, 75-86, 1996.

Webb, S.L. and D.B. Dingwell, Non-Newtonian rheology of igneous melts at high stresses and strain rates: Experimental results for rhyolite, andesite, basalt and nephelinite, Scarfe Volume, *J. Geophys. Res.*, **95**, 15695-15701, 1990a.

Webb, SL. and D.B. Dingwell, The onset of non-Newtonian rheology of silicate melts: A fiber elongation study, *Phys. Chem. Mineral.*, **17**, 125-132, 1990b.

Webb, S.L. and D.B. Dingwell, Compressibility of titanosilicate melts, *Contrib. Mineral. Petrol.*, **118**, 157-168, 1994.

Webb, S.L. and D.B. Dingwell, Viscoelasticity, In: Structure, Dynamics and Properties of Silicate Melts, J.F. Stebbins, P.F. McMillan and D.B. Dingwell (eds), *Rev. Mineral.*, **32**, 95-119, 1995.

Webb, S.L. and R. Knoche, The glass-transition, structural relaxation and shear viscosity of silicate melts, *Chem. Geol.*, **128**, 165-183, 1996.

Webb, S.L., R. Knoche and D.B. Dingwell, Determination of silicate liquid thermal expansivity using dilatometry and calorimetry, *Europ. J. Mineral.*, **4**, 95-104, 1992.

Weill, D.F., R. Hon and A. Navrotsky, The igneous system $CaMgSi_2O_6$-$CaAl_2Si_2O_8$-$NaAlSi_3O_8$: variations on a classic theme by Bowen, In: Physics of Magmatic Processes, R.B. Hargraves (ed), Princeton University Press, 49-92, 1980.

Wilding, M., S.L. Webb and D.B. Dingwell, Evaluation of a relaxation geospeedometer for volcanic glasses, *Chem. Geol.*, **125**, 137-148, 1995.

Wilding, M., S.L. Webb and D.B. Dingwell, Tektite cooling rate - calorimetric relaxation geospeedometry applied to a natural glass, *Geochim. Cosmochim. Acta*, **60**, 1099-1103, 1996.

Wong, J. and C.A. Angell, Glass Structure by Spectroscopy, New York, 864p, 1976.

Zarzycki, J., Glasses and the Vitreous State, Cambridge University Press, Cambridge, 505p, 1991.

Zdaniewski, W.A., G.E. Rindone and D.E. Day, The internal friction of glasses, *J. Mater. Sci.*, **14**, 763-775, 1979.

Index

Lecture Notes in Earth Sciences